The Bats of Texas

Number Eleven
The W. L. Moody, Jr., Natural History Series

The Bats of Texas

DAVID J. SCHMIDLY

Drawings by Christine Stetter

TEXAS A&M UNIVERSITY PRESS
COLLEGE STATION

Second printing, 1999

The paper used in this book meets the minimum requirements
of the American National Standard for Permanence
of Paper for Printed Library Materials, Z39.48-1984.
Binding materials have been chosen for durability.

LIBRARY OF CONGRESS CATALOGING-IN-PUBLICATION DATA
Schmidly, David J., 1943–
The bats of Texas / David J. Schmidly ; drawings by Christine Stretter. – 1st ed.
p. cm. – (The W.L. Moody, Jr., natural history series ; no. 11)
Includes bibliographical references and index.
ISBN 0-89096-403-3 (alk. paper). –
ISBN 0-89096-450-5 (pbk. : alk. paper)
1. Bats–Texas. I. Title. II. Series.
QL737.C5S344 1991
599.4′09764–dc20 90-39108
CIP

In memory of my father,
H. J. (Chick) Schmidly,
in recognition of his support
and friendship

Contents

Illustrations

FIGURES

COLOR PLATES

Tables

Preface

Bats are among nature's most misunderstood animals. No other group of mammals has been so shrouded in mystery, myth, and misinformation as bats. They abound in mythology and folklore, normally portrayed as sinister, demoniac, and generally undesirable creatures. The German word for a bat, *fledermaus*, translates to "flying mouse."

Bats suffer from a bad public image. Many people regard them as blind, dirty and dangerous, common carriers of rabies, and ugly creatures that try to become entangled in human hair. In fact, however, bats are not blind (they see quite well, especially in the dark), they don't fly into people's hair or make them go crazy, and they are not even remotely related to mice.

Bats are intelligent, extremely interesting, and possess fascinating abilities. Not only can they fly, which is unique among mammals, but they can do so in complete darkness—and negotiate obstacles and catch insects at the same time. Their diet is varied and includes insects, pollen, fruits, flowers, flesh, and blood. Their social organization ranges from solitary through small family groups to harems, to immense colonies of several million individuals. Some species are able to slow their body processes down to an absolute minimum in order to survive cold temperatures. Others are known to migrate hundreds or even thousands of miles in search of suitable living conditions.

Bats have been the subject of increased interest and study in the United States in recent years primarily because of their intriguing, and in many ways unique, biological properties, but also because some species have been associated with diseases affecting people (rabies and histoplasmosis), and because some bats are highly beneficial to us. Frugivorous and nectarivorous bats are important pollinators and seed dispersers for an array of useful plants. Insectivorous bats are practically the only predators of night-flying insects and are responsible for destroying tons of such pests annually. Bat guano is collected for fertilizer, and bats are also valuable as study animals for scientific research.

The present treatment is a synopsis of current knowledge about Texas bats. There are four families and thirty-two species of bats in Texas. Our bat fauna includes all of the families and all but nine of the species that occur in the United States. Although five of these species are represented by only one specimen and may be regarded as vagrants, no other state has a bat fauna as diverse as that of Texas.

The ultimate aim of this book is to stimulate interest and provide better understanding about these often maligned mammals, and to promote greater appreciation of their role in our biological communities and of the need to conserve them. Davis's (1974) section on bats in his *Mammals of Texas* is the most recent publication in which the distribution and biology of bats in the state were treated in detail. The present book is intended to serve as a single reference that summarizes new information, and integrates it with that already available in published form, on the systematics, distribution, and biology of Texas bats.

The first chapter is about bats in general, providing basic information about their appearance, distribution, classification, evolution, biology, and life history and about public health and bat conservation. However, it is not a comprehensive treatise on bat biology. The second chapter contains a dichotomous key, with illustrations, to aid in the identification of Texas bats. The third, and most extensive, chapter presents a synopsis of current knowledge for the thirty-two species of bats known to occur in our state. Finally, the fourth chapter contains a detailed bibliography and list of references for bats in Texas, other parts of the United States, and northern Mexico. All of the literature cited in the text has been listed, as well as several other general references and technical reports for the interested reader seeking greater detail.

Accounts for each species have been arranged so that they contain in sequence (1) a brief description of the bat, accompanied by a photograph, with special emphasis given to distinguishing features and comparisons to similar species; (2) a description of the distribution of the species in Texas, with reference to a map; (3) the name and appropriate scientific authority for all the subspecies of a particular bat in Texas as well as the most recent taxonomic treatment of the species; (4) a discussion of the animal's life history, including habitat/roosting preferences, reproduction, and food habits; (5) a list of Texas localities where scientific specimens of bats have been obtained, including the number of specimens collected and the acronym for the museum or collection where the specimens are housed; and (6) a comprehensive listing of published material about each species, indicated by numbers corresponding to the numbered citations provided in chapter four. The life histories include observations recorded by other researchers and reported in the literature as well as my personal experiences based on more than twenty-five years of field work in Texas.

Texas exhibits a wide range of climate, landforms, and vegetation which combine to produce a diverse state geography. Ten ecological regions have been identified and described in the state (fig. 1), and these are used to describe the ranges of bats in the state. The interested reader is referred to Gould (1975) for a detailed description of each of the major ecological regions.

Altogether I have examined more than six thousand specimens of bats from Texas that have been collected by professional mammalogists and deposited in various collections throughout the United States. In preparing distribution maps I have taken into account reports on bats from adjacent states in the United States, including Lowery (1974) for Louisiana, Harvey (1986) and Selander (1979)

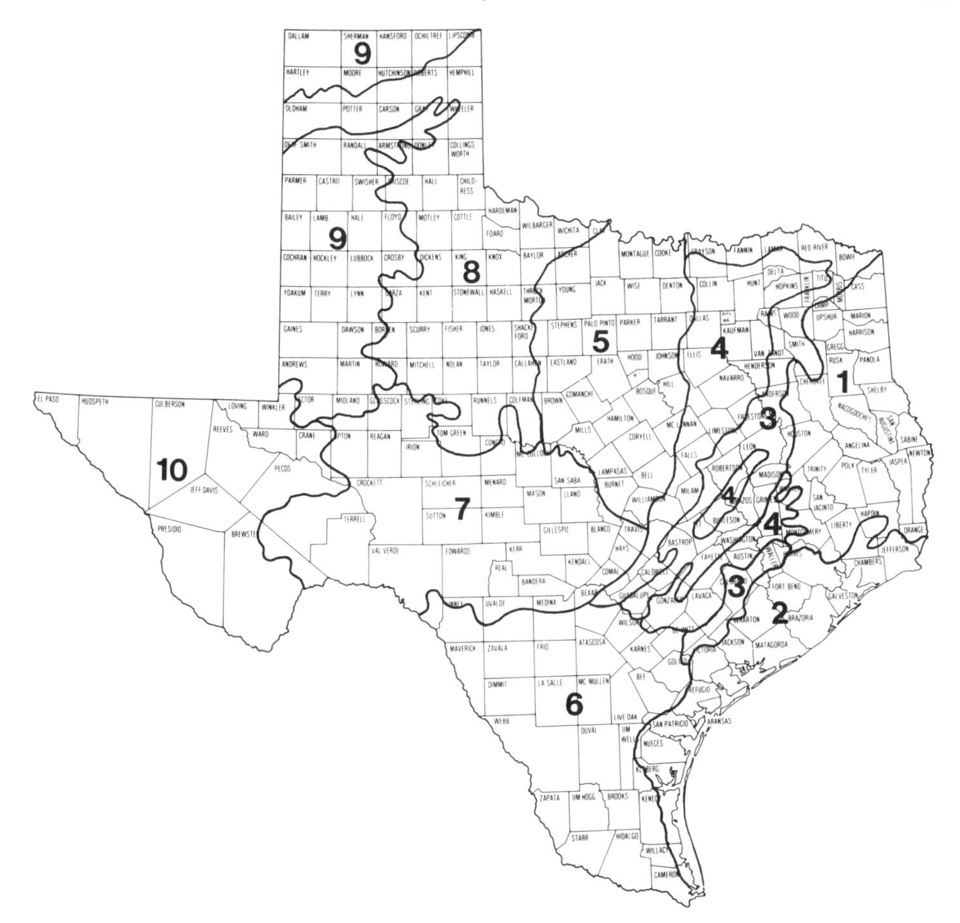

Figure 1. Ecological regions of Texas. (*Source:* F. W. Gould. 1962. *Texas plants: a checklist and ecological summary.* Texas Agricultural Experiment Station. MP-585.)

for Arkansas, Findley et al. (1975) and Findley (1987) for New Mexico, and Hoffmeister (1986) for Arizona, as well as reports from those Mexican states which border Texas, namely Alvarez (1963) for Tamaulipas, Baker (1956) for Coahuila, and Anderson (1972) for Chihuahua.

On distribution maps, the outlined area is my estimate of the probable distribution in Texas of the species concerned; if no heavy lines are used anywhere on the map, the species is to be expected (in suitable habitats) in any part of the state. On these maps, localities represented by specimens I have examined are indicated by black dots whereas those based on literature records are indicated by black squares. Black triangles represent bats reported to the Texas Department of Health, and which I later examined and identified.

A list of specimens examined is included at the close of each species account. Although measures given in other sections of the book are in metric, those listed in specimens examined are in Imperial measure, as they were originally recorded. Acronyms used for the various collections (see table 1) follow Yates et al. (1987) and are listed in the acknowledgements. Other abbreviations periodically used in these listings include the following: BBNP (Big Bend National Park), BGWMA

Table 1. Museum collections containing Texas bat specimens, and acronyms used to identify the various collections in the *Specimens examined* section of each species account

Acronym	Collection
AMNH	American Museum of Natural History, New York, N.Y.
ANSP	Philadelphia Academy of Natural Sciences, Philadelphia, Pa.
ASVRC	Angelo State University Natural History Collection, San Angelo, Tex.
BBNHA	Big Bend Natural History Association, Big Bend National Park, Tex.
CCSU	Corpus Christi State University Vertebrate Collection, Corpus Christi, Tex.
CM	Carnegie Museum of Natural History, Pittsburgh, Pa.
DMNHT	Dallas Museum of Natural History, Dallas, Tex.
FMNH	Field Museum of Natural History, Chicago, Ill.
FWMSH	Fort Worth Museum of Science and History, Fort Worth, Tex.
KU	University of Kansas, Museum of Natural History, Lawrence, Kans.
LACM	Natural History Museum of Los Angeles County, Los Angeles, Calif.
LSUMZ	Louisiana State University, Museum of Zoology, Baton Rouge, La.
MSB	Museum of Southwestern Biology, University of New Mexico, Albuquerque, N. Mex.
MSU	The Museum, Michigan State University, East Lansing, Mich.
MVZ	Museum of Vertebrate Zoology, University of California–Berkeley, Berkeley, Calif.
MWSU	Midwestern State University Collection of Recent Mammals, Wichita Falls, Tex.
NTSU	North Texas State University, Department of Biological Sciences, Denton, Tex.
NWMSU	Northwest Missouri State University, Department of Biology, Maryville, Mo.
SFASU	Stephen F. Austin University, Department of Biology, Nacogdoches, Tex.
SM	George W. Carroll Mammal Collection, Strecker Museum, Baylor University, Waco, Tex.
SRSU	Vertebrate Collection, Sul Ross State University, Alpine, Tex.
TAIU	Texas A&I Collections, Texas A&I University, Kingsville, Tex.
TCWC	Texas Cooperative Wildlife Collection, Texas A&M University, College Station, Tex.
TNHC	Texas Natural History Collection, Texas Memorial Museum, University of Texas at Austin, Austin, Tex.
TTU	The Museum, Texas Tech University, Lubbock, Tex.
TWC	Texas Wesleyan College, Museum of Zoology, Fort Worth, Tex.
UIMNH	University of Illinois, Museum of Natural History, Urbana, Ill.
UMMZ	Mammal Collections, Museum of Zoology, University of Michigan–Ann Arbor, Ann Arbor, Mich.
USNM/FWS	National Museum of Natural History, Vertebrate Zoology Department/ U.S. Fish and Wildlife Service, Washington, D.C.
UTACV	UTA Collection of Vertebrates, University of Texas at Arlington, Arlington, Tex.
UTEP	Mammal Division, Resource Collections, Laboratory for Environmental Biology, University of Texas at El Paso, El Paso, Tex.
WMM	Witte Memorial Museum, San Antonio, Tex.

(Black Gap Wildlife Management Area), DMSP (Davis Mountain State Park), GMNP (Guadalupe Mountains National Park), and BTNP (Big Thicket National Preserve). Collecting localities recorded in the literature for which I did not examine the specimens are listed separately as additional records, along with the appropriate literature citation.

Common and scientific names of Texas bats are used following Jones et al. (1986, 1988), except where recent taxonomic study has required changes. As is customary, the authority first describing a species, and the date of description, are named following the Latin name of the bat. If, subsequent to the original description, the bat has been placed in a different genus, the describer's name and the date are placed in parentheses. This style of reporting names should not be confused with literature citations. Because of recent taxonomic revisions based on study of new material or the use of new technologies, only seven Texas bats have retained their original generic names, with the describer listed without parentheses.

Acknowledgments

Completion of this book would not have been possible without the cooperation and assistance of a number of people. I am especially grateful to those in charge of the collections at the numerous institutions where I examined specimens; these institutions are listed in table 1.

Photographs of each bat have been included to help in identification, and to put a face on the subject. The photographs were provided by John Tveten, a nature photographer from Baytown, Texas, who has worked with me on numerous projects; Dr. Merlin Tuttle, who is the founder and science director of Bat Conservation International in Austin, Texas; Dr. Scott Altenbach of the Department of Biology at the University of New Mexico; Dr. Bruce Hayward of the Department of Biology at Western New Mexico University; and Dr. Roger W. Barbour, distinguished professor emeritus, School of Biology, University of Kentucky. The cooperation and assistance of these individuals is gratefully acknowledged and warmly appreciated.

David Scarbrough assisted with library work, listed specimens examined, prepared distribution maps, and drafted earlier versions of the species accounts. His ability to organize material and his attention to detail were invaluable throughout the entire project. Christine Stetter prepared the illustrations for the key and general anatomy of bats. Joe Wise and Vic Whadford of the Texas Department of Health carefully noted information on and preserved bats shipped to their office, and then forwarded these specimens to our laboratories so that they could be identified properly and have additional data recorded. Paisley Cato and George Baumgardner sorted and assisted with the identification of these bats. Prilla Tucker visited mammal collections throughout Texas to record the collection date, location, and other available natural history documentation for each bat specimen. Her summary of this information contributed substantially to this project.

The Caesar Kleberg Foundation and the Texas Agricultural Experiment Station graciously provided financial support for some of the field work and all the trips to museums and collections.

The Bats of Texas

1

Introduction

GENERAL APPEARANCE

Bats are mammals and, as such, possess all the features characteristic of this vertebrate class, including a body covering of hair (pelage or fur) and mammary glands for the production of milk to suckle and nourish their newborn young. Bats differ from other mammals, however, in possessing wings that flap—a character which, coupled with their nocturnal habits, has played a prominent role in their association with folklore and superstition. Even so-called flying mammals, such as flying squirrels and flying lemurs (or colugos), which possess expanded flaps of skin, are not able to undertake powered flight—they merely glide.

Bats are so highly specialized for flight (fig. 2) that locomotion by other means is accomplished with difficulty. The wing pattern is essentially similar in all species, but differences in shape reflect the variety of ecological niches and feeding behaviors exhibited by bats. The upper arm, or humerus, is shorter than the forearm, which is composed of a thin, thread-like ulna fused to a longer, somewhat curved radius. All bats have a short, clawed thumb, used mainly for moving around roosts, and four greatly elongated fingers that serve to spread and manipulate the wing. The thumb also serves as an attachment for the propatagium, which is that part of the wing membrane in front of the forearm.

The hind limbs of bats are relatively small and attached at the hip in a reverse manner from those of other mammals. Thus, the knee is directed outward due to a 90° rotation of the hind leg, and the foot has rotated 180° from the usual mammalian position so that it points backward and facilitates a head-down suspension of the animal. The hind foot has five toes of approximately equal length, all bearing well-developed curved claws which bats use to hang by their toes.

The wing membrane, called the patagium, consists of an upper and lower skin layer, between which are sandwiched bundles of elastic tissue and muscle fiber. This structure is attached along the sides of the body and hind legs and is braced by the elongated finger bones, arms, and legs. Four distinct flight membranes are recognized. The propatagium extends from the shoulder to the wrist anterior to the upper arm and forearm. The chiropatagium is that portion of the wing membrane forming an elastic webbing between the fingers of the hand. The plagiopatagium extends from digit five to the side of the body and the hind leg.

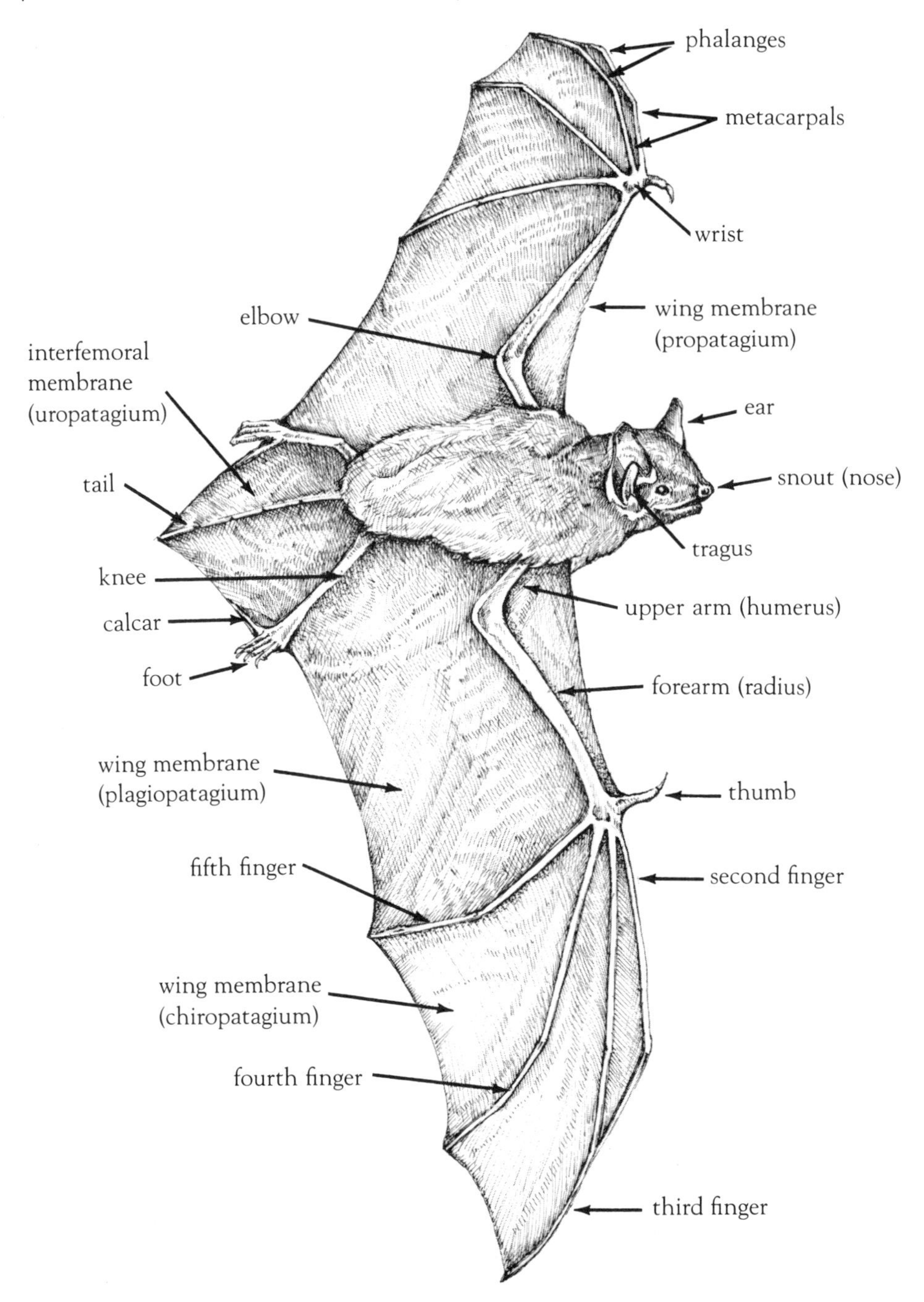

Figure 2. Anatomy of a typical bat labeled to show names of parts, as used in text

Finally, the area of the wing membrane between the hind limbs and the tail is called the uropatagium or interfemoral membrane. The tail membrane is further supported in many bats by a long cartilaginous spur, called the calcar, which articulates with the heel of the foot. Tail membranes help increase aerial maneuverability and serve as scoops to assist insect-eating bats in catching prey in midair, as pouches to receive baby bats during delivery, and as blankets to help foliage-roosting bats conserve body heat.

Bats exhibit a wide variety of head shapes, which is primarily a reflection of their wide variation in diet and method of food capture (fig. 3). Faces of many species look peculiar because of their oddly shaped and often enlarged ears; leaf-like projections and other elaborate structures on or around their noses; and wrinkles, lumps, and bumps on their lips. Among the most curious of head shapes in Texas bats are those found in the nectar and pollen feeders, in which the snout is greatly elongated and the back of the head is low and rounded. Although not present in all species of bats, an interesting structure called the tragus exists in the ears of all Texas bats. The tragus is a thin, erect, fleshy projection rising from the inner base of the ear. This structure varies considerably in size and shape and is occasionally used for taxonomic purposes, but the functional significance of the tragus is not easily explained.

Worldwide, bats vary substantially in size, ranging from the world's smallest mammal, Kitti's hog-nosed bat (wingspan, 130–45 mm; weight, 1.5 g) which lives in Thailand and is about the size of a large bumblebee, to the large "flying foxes" of Africa, Asia, Australia, and many Pacific islands with a reported weight of 1.5 kg and a wingspan of nearly two meters. Such size variations are less evident in Texas bats, with the smallest species (*Pipistrellus hesperus*) having a wingspan of 227 mm and the largest (*Eumops perotis*) 570 mm (table 2).

Nearly all Texas bats have a tail, although as with many of the structures discussed above, there is considerable variation (fig. 3). The bats of the family Molossidae are all characterized by having at least half of the tail projecting beyond the rear margin of the uropatagium, which is where they get the common name "free-tailed" bats. Bats of the family Vespertilionidae have long tails that are completely bound within the uropatagium. In the family Mormoopidae, the tail protrudes for about 10–15 mm from the top surface of the interfemoral membrane at about the level of the knee. The two species of the family Phyllostomatidae that occur in Texas either lack a tail altogether or, if present, it is very short (less than 10 mm in length).

In common with other nocturnal or crepuscular mammals bats have, as a rule, much more sober coloring than diurnal mammals. Texas bats are mostly colored drab shades of brown, black, and gray, with the ventral surface a lighter tint of the same color that pervades the back. Four species in our state, however, diverge in color from the quiet tones of their relatives. These are the spotted bat (*Euderma maculatum*), which is black on the dorsal surface, with three large white patches and enormous pink ears (fig. 4); the red bat (*Lasiurus borealis*) and the hoary bat (*L. cinereus*), which have variegated fur tipped with white (fig. 4); and the silver-haired bat (*Lasionycteris noctivagans*) which has a white frosting on the fur that gives it a grizzled appearance (fig. 4).

The permanent teeth of bats, as in most other mammals, consist of four kinds: incisors, canines, premolars, and molars. All except the molars are deciduous, meaning they are replaced once in the life span of the individuals. The milk teeth differ in both number and form from the permanent set. They are tiny, sharp-

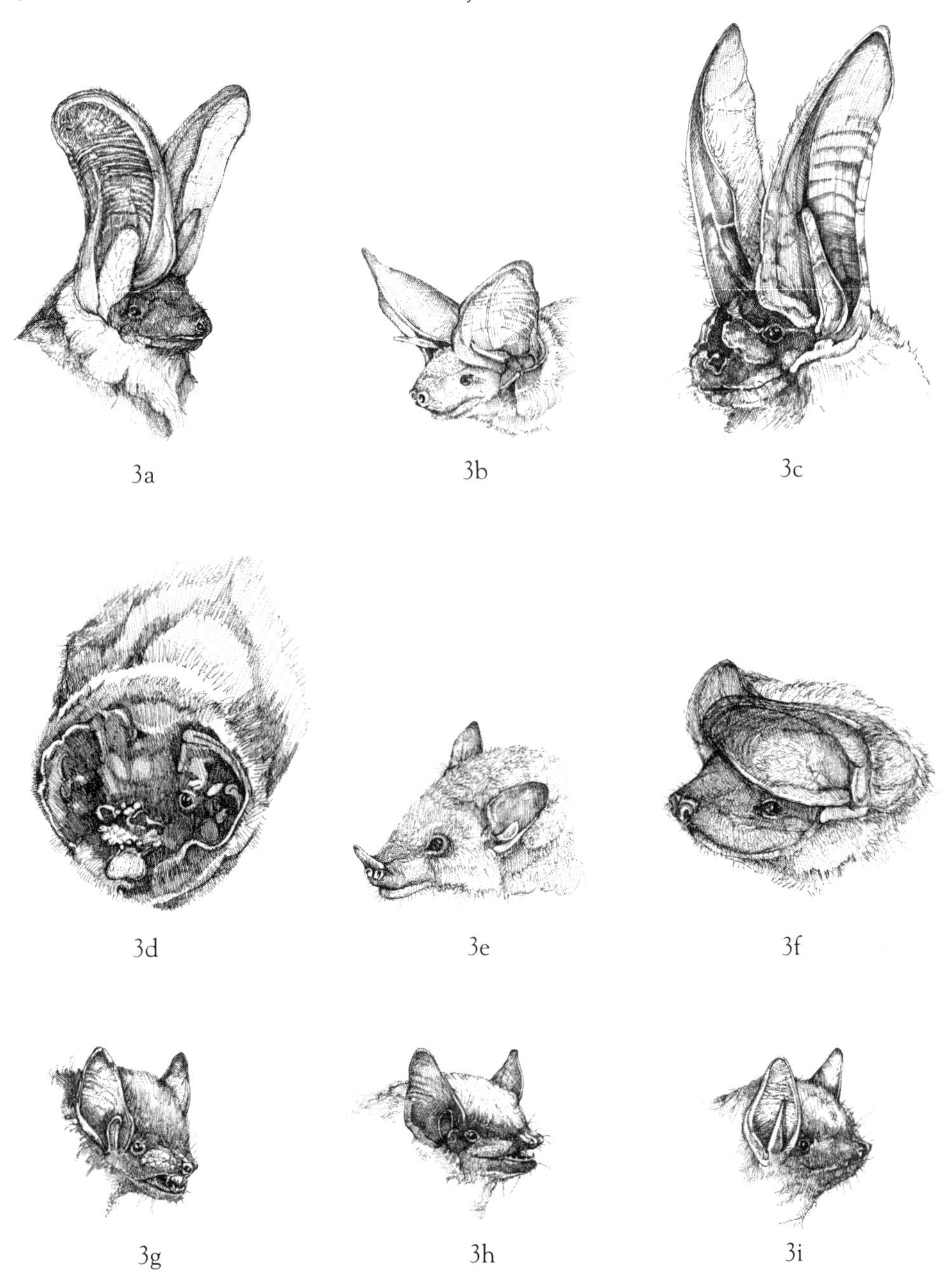

Figure 3. Facial portraits of several Texas bats: (a) *Euderma maculatum*, (b) *Antrozous pallidus*, (c) *Plecotus townsendii*, (d) *Mormoops megalophylla*, (e) *Leptonycteris nivalis*, (f) *Eumops perotis*, (g) *Eptesicus fuscus*, (h) *Pipistrellus hesperus*, and (i) *Myotis velifer*

pointed, and hooked to enable the young bat to cling more effectively to the teat of its mother while she is carrying her offspring in flight. The permanent teeth differ widely among genera and species, again in both number and form (table 3). The greatest number of teeth present in a Texas bat is thirty-eight, which is the number in all species of the genus *Myotis*. The least number of teeth found

TABLE 2. Average body weights and wing measurements for Texas bats

Species	Weight (grams)	Forearm length (mm)	Wingspan (mm)	Wing area (cm^2)
M. megalophylla	13–19	51–59	370	188
C. mexicana	10–25	43–45	345	166
L. nivalis	24	55–60	410	223
D. ecaudata	30–40	50–56	326	110
M. lucifugus	7–9	34–41	239	86
M. yumanensis	4–6	32–38	225	77
M. austroriparius	5–7	36–41	254	104
M. velifer	15	37–47	296	131
M. septentrionalis	5–9	32–39	241	91
M. thysanodes	6–11	39–46	285	125
M. volans	5–9	35–41	267	108
M. californicus	3–5	29–36	220	77
M. ciliolabrum	4–5	30–36	242	83
L. noctivagans	8–12	37–44	289	104
P. hesperus	3–6	27–33	190	60
P. subflavus	4–6	31–35	237	81
E. fuscus	13–20	42–51	325	162
L. borealis	10–15	35–45	312	134
L. blossevillii	10–15	39–42	295	110
L. seminolus	10–15	35–45	300	104
L. cinereus	20–35	46–58	400	180
L. intermedius	18–24	45–56	370	203
L. ega	10–15	45–48	345	148
N. humeralis	5–7	33–39	263	107
E. maculatum	16–20	48–51	365	181
P. townsendii	7–12	39–48	293	136
P. rafinesquii	7–13	40–46	270	143
A. pallidus	12–17	48–60	353	228
T. brasiliensis	11–14	36–46	301	110
N. femorosacca	10–14	44–50	345	132
N. macrotis	24–30	58–64	426	178
E. perotis	65	72–82	550	322

in any bat in our state is twenty-six, in the hairy-legged vampire bat (*Diphylla ecaudata*).

DISTRIBUTION

Bats live nearly everywhere on the earth with the exception of the polar regions, highest mountains, and some remote islands. However, their diversity and abundance is greatest in tropical regions, declining steadily north and south of the equator. Bats are common in the United States and can be found easily in most regions, although they are most abundant in the Southwest.

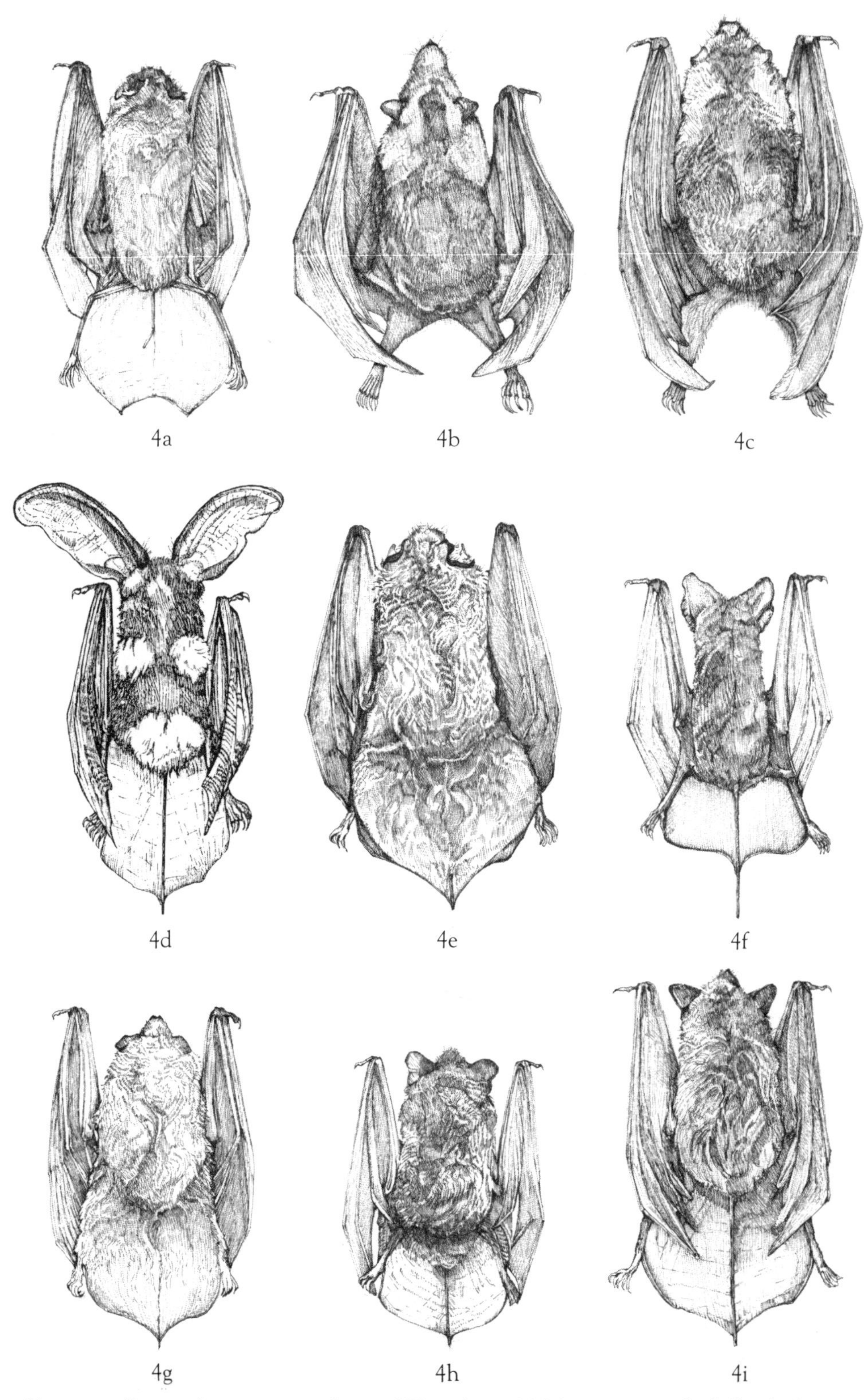

Figure 4. External appearance of several Texas bats: (a) *Mormoops megalophylla*, (b) *Leptonycteris nivalis*, (c) *Diphylla ecaudata*, (d) *Euderma maculatum*, (e) *Lasiurus cinereus*, (f) *Tadarida brasiliensis*, (g) *Lasiurus borealis*, (h) *Lasionycteris noctivagans*, and (i) *Eptesicus fuscus*

TABLE 3. Dental formulas for Texas bats

Species	Upper teeth[1] I	C	Pm	M[2]	Lower teeth[1] I	C	Pm	M[2]	Total (×2)
M. megalophylla	2	1	2	3	2	1	3	3	34
C. mexicana	2	1	2	3	0	1	3	3	30
L. nivalis	2	1	2	2	2	1	3	2	30
D. ecaudata	2	1	1	2	2	1	2	2	26
M. lucifugus	2	1	3	3	3	1	3	3	38
M. yumanensis	2	1	3	3	3	1	3	3	38
M. austroriparius	2	1	3	3	3	1	3	3	38
M. velifer	2	1	3	3	3	1	3	3	38
M. septentrionalis	2	1	3	3	3	1	3	3	38
M. thysanodes	2	1	3	3	3	1	3	3	38
M. volans	2	1	3	3	3	1	3	3	38
M. californicus	2	1	3	3	3	1	3	3	38
M. ciliolabrum	2	1	3	3	3	1	3	3	38
L. noctivagans	2	1	2	3	3	1	3	3	36
P. hesperus	2	1	2	3	3	1	2	3	34
P. subflavus	2	1	2	3	3	1	2	3	34
E. fuscus	2	1	1	3	3	1	2	3	32
L. borealis	1	1	2	3	3	1	2	3	32
L. blossevillii	1	1	2	3	3	1	2	3	32
L. seminolus	1	1	2	3	3	1	2	3	32
L. cinereus	1	1	2	3	3	1	2	3	32
L. intermedius	1	1	1	3	3	1	2	3	30
L. ega	1	1	1	3	3	1	2	3	30
N. humeralis	1	1	1	3	3	1	2	3	30
E. maculatum	2	1	2	3	3	1	2	3	34
P. townsendii	2	1	2	3	3	1	3	3	36
P. rafinesquii	2	1	2	3	3	1	3	3	36
A. pallidus	1	1	1	3	2	1	2	3	28
T. brasiliensis	1	1	2	3	2/3	1	2	3	30/32
N. femorosacca	1	1	2	3	2	1	2	3	30
N. macrotis	1	1	2	3	2	1	2	3	30
E. perotis	1	1	2	3	2	1	2	3	30

[1]Number of teeth in *each side* of jaw.
[2]I=Incisors; C=Canines; Pm=Premolars; M=Molars

Bats occur in all the major ecological regions of Texas (table 4). Of the four major environmental components that determine their distribution and abundance (climate, roosts, food, and other animals—predators and competitors), the first and second of these factors probably are most important. Locations characterized by a high degree of topographic relief, indicative of a highly variable or patchy environment, typically support a high density and abundance of bats, primarily because these areas have a greater number of roosting sites and a greater diversity of food resources for bats to exploit.

The Big Bend region of the Trans-Pecos, with its topographic pattern of high mountains and desert lowlands, supports more kinds of bats (seventeen species) than any other part of Texas. Several extremely rare or unusual bats occur in this

Table 4. Distribution of Texas bats according to the ten ecological regions of Texas

Species	Pineywoods	Gulf Coast Prairies & Marshes	Post Oak Savannah	Blackland Prairies	Cross Timbers and Prairies	South Texas Plains	Edwards Plateau	Rolling Plains	High Plains	Trans-Pecos
M. megalophylla						X	X			X
C. mexicana						X				
L. nivalis										X
D. ecaudata							X			
M. lucifugus										X
M. yumanensis						X	X			X
M. austroriparius	X	X								
M. velifer			X	X	X	X	X	X	X	X
M. septentrionalis						X				
M. thysanodes									X	X
M. volans								X		X
M. californicus										X
M. ciliolabrum								X	X	X
L. noctivagans	X	X					X	X	X	X
P. hesperus							X	X	X	X
P. subflavus	X	X	X	X	X	X	X	X		
E. fuscus	X	X		X	X			X	X	X
L. borealis	X	X	X	X	X	X	X	X	X	X
L. blossevillii										X
L. seminolus	X	X	X							
L. cinereus	X	X	X	X	X	X	X	X	X	X
L. intermedius	X	X	X	X		X	X			
L. ega		X				X				
N. humeralis	X	X	X	X	X	X	X			
E. maculatum										X
P. townsendii							X	X	X	X
P. rafinesquii	X									
A. pallidus						X	X	X	X	X
T. brasiliensis	X	X	X	X	X	X	X	X	X	X
N. femorosacca										X
N. macrotis		X	X						X	X
E. perotis							X			X
Total	11	12	9	8	7	13	15	12	12	22

region, and in an abundance known nowhere else in the country. The Edwards Plateau also maintains a high diversity of bats, primarily cavern-dwelling species that inhabit the numerous caves of this region, often in staggering numbers. Caves are excellent roosting and hibernation sites for bats, and they play a prominent role in the distribution of many bats in our state. Bat species diversity and numbers are lower in the northern, eastern, and southern portions of the state, where topographic heterogeneity is low and caves are uncommon.

In many parts of North America, bats either migrate or hibernate in winter. If they migrate, their distributions in winter and summer are often very different (table 5). In parts of Texas where the climate is mild, bats may not migrate or hibernate if weather conditions are such that a sufficient and suitable food supply is available year-round. In general, however, there is a tendency for most species of bats in our state either to move from one region to another or to move out of the state altogether during the months of November through March.

CLASSIFICATION

Bats belong to the mammalian order Chiroptera, which means "hand-wing." The order is divided into two suborders, the Megachiroptera and the Microchiroptera. The former is distinguished from the latter by the fact that most of its genera have a claw on the second digit, whereas in the Microchiroptera only the thumb bears a claw.

The Chiroptera is the largest mammalian order after the rodents, with 17 families, 175 genera, and approximately 919 species. The Megachiroptera comprises only 1 family with 161 species distributed in the tropical and subtropical regions of the Old World. The suborder Microchiroptera contains 16 families, 758 species, and is cosmopolitan.

EVOLUTION

Bats are thought to have originated in the Paleocene or mid to late Cretaceous period of prehistory (some 70 to 100 million years ago), but their fossil record is poor. Approximately seventy genera (both living and extinct) are known, with the earliest fossils dating back to the early Eocene (approximately 60 million years ago). Many of these genera are represented only by fragmentary and incomplete fossils but some of the earliest known bats are quite well preserved. Even these, however, are well developed and modern in appearance; therefore, transitional links between modern bats and their progenitors have yet to be discovered. Although fossils of primitive bats are still unknown, it has been suggested that bats may have evolved from small, arboreal insectivores with gliding abilities (Hill and Smith, 1984). The available evidence indicates that bats quickly evolved into their modern form and then rapidly specialized to become the large and diverse order of mammals seen today.

TABLE 5. Seasonal occurrence of bats in Texas

Species	Month											
	J	F	M	A	M	J	J	A	S	O	N	D
M. megalophylla	—	—	—		—	—	—	—	—	—	—	—
C. mexicana												—
L. nivalis						—	—	—				
D. ecaudata					—							
M. lucifugus					—	—						
M. yumanensis			—	—	—	—	—	—	—	—		
M. austroriparius		—	—	—	—	—	—			—		—
M. velifer	—	—	—	—	—	—	—	—	—	—	—	—
M. septentrionalis								—				
M. thysanodes				—	—	—	—	—				
M. volans				—	—	—	—	—	—			
M. californicus	—	—	—	—	—	—	—	—	—	—		
M. ciliolabrum			—	—	—	—	—					
L. noctivagans	—		—	—	—	—			—	—	—	—
P. hesperus	—	—	—	—	—	—	—	—	—			
P. subflavus	—	—	—	—	—	—	—	—	—	—	—	—
E. fuscus	—	—	—	—	—	—	—	—	—	—	—	—
L. borealis	—	—	—	—	—	—	—	—	—	—	—	—
L. blossevillii							—					
L. seminolus	—	—	—	—	—	—	—	—	—	—		
L. cinereus	—	—	—	—	—	—	—	—	—	—		
L. intermedius	—	—	—	—	—	—	—	—	—	—		
L. ega					—	—			—			
N. humeralis	—	—	—	—	—	—	—	—	—	—	—	—
E. maculatum						—	—	—				
P. townsendii	—	—	—	—	—	—	—	—	—	—	—	—
P. rafinesquii				—	—	—				—	—	—
A. pallidus			—	—	—	—	—	—	—	—		
T. brasiliensis	—	—	—	—	—	—	—	—	—	—	—	—
N. femorosacca						—	—	—				
N. macrotis					—	—	—	—		—	—	
E. perotis	—		—			—	—	—	—			

TABLE 6. Fossil bats from Texas

Species	Reference(s)
*Macrotus californicus**	Cockerell (1930), Ray and Wilson (1979)
*Desmodus rotundus**	Cockerell (1930), Ray and Wilson (1979)
Myotis lucifugus	Dalquest and Stangl (1984b), Roth (1972)
Myotis velifer	Choate and Hall (1967), Dalquest and Stangl (1984a), Dalquest and Stangl (1984b), Dalquest et al. (1969), Dorsey (1977), Logan and Black (1979), Lundelius (1967), Patton (1963)
Myotis volans	Roth (1972)
*Myotis evotis**	Dalquest et al. (1969)
Myotis californicus	Roth (1972)
Myotis rectidentis†	Choate and Hall (1967), Dorsey (1977)
Myotis sp.	Semken (1961)
Lasionycteris noctivagans	Dalquest (1978), Logan and Black (1979)
Pipistrellus hesperus	Roth (1972)
Pipistrellus subflavus	Dalquest et al. (1969)
near *Eptesicus fuscus*	Dalquest (1978)
Eptesicus fuscus	Dalquest and Carpenter (1988), Dalquest and Stangl (1984b), Dalquest et al. (1969), Logan and Black (1979), Patton (1963), Slaughter and McClure (1965)
Eptesicus hemphillensis†	Dalquest (1983)
Lasiurus borealis	Dalquest (1978)
Lasiurus cinereus	Dalquest et al. (1969)
Lasiurus sp.	Dalquest (1967)
Plecotus townsendii	Logan and Black (1979)
Pizonyx wheeleri†	Dalquest and Patrick (1989)
Antrozous pallidus	Dalquest (1978), Logan and Black (1979), Van Devender et al. (1987)
near *Tadarida*	Dalquest (1975)

*No longer occurs in Texas
†Extinct

Fossil bats from Texas (table 6) are primarily of living species still occurring in the state. However three extinct bats are recorded from Texas (*Myotis rectidentis*, *Eptesicus hemphillensis*, and *Pizonyx wheeleri*), and three species (*Macrotus californicus*, *Desmodus rotundus*, and *Myotis evotis*) for which there are fossil records no longer occur in the state. *Macrotus californicus* is a big-eared phyllostomatid found in the southwestern United States and Mexico and *Desmodus rotundus*, the common vampire, ranges throughout Mexico. *Myotis evotis* is a common bat of western and northwestern North America.

FLIGHT

Obviously, the central feature in the evolution of bats and the most diagnostic feature of these mammals is the development of wings and their ability to fly. The power of flight has allowed bats to exploit a large food base unavailable to terrestrial mammals (flying insects), reduces potential predation, and allows bats to exploit large geographic ranges. These abilities, due to the power of flight, are responsible for the great diversity and abundance of bats worldwide.

Compared with the wings of birds, bat wings are simple in structure. The wing skeleton contains all the bones evident in the forelimbs of most mammals (fig. 2) and, in fact, is merely an elongated hand. A flexible membrane, or patagium, is stretched between the body and the fingers of this hand to complete the structure of the wing. The curvature of the entire structure (camber) can be altered by the bat at will to increase or decrease lift.

Nine muscles or muscle groups are involved in bat flight. Four large muscles attach to the upper arm (humerus) and scapula in bats and are responsible for the downstroke, or powerstroke. Eight smaller muscles act on the upstroke of the bat wing. The powerstroke muscles are located in the chest and the upstroke muscles on the back. This arrangement differs markedly from the flight muscles of birds, which have only one muscle responsible for the powerstroke and one for the upstroke. In birds, both flight muscles are located in the chest.

The specialization of bats for different feeding strategies and habits is often evident in the shape and size of their wings. The aspect ratio (calculated as the wingspan squared, divided by the area of the wing) is a common measurement used to express functional characteristics of bat wings (Farney and Fleharty, 1969), and it has been demonstrated that flight speed is positively correlated with this trait (Findley et al., 1972). High aspect ratios denote a narrow wing, low aspect ratios a broad one. Increasing aspect ratio decreases drag, thus permitting greater speed, but at the same time reduces lift. Low aspect ratio wings generate considrable drag at higher speeds, but also promote maximal lift at low speeds. From these statements, it would be expected that long, narrow wings (with high aspect ratios) are indicative of rapid and/or sustained flight, whereas short, wide wings (with low aspect ratios) indicate slow and maneuverable flight (Struhsaker, 1961).

Comparisons of wing measurements for Texas bats are provided in table 2. The species in the family Molossidae have high aspect ratios with long, very narrow wings. These bats are capable of swift flight and are known to forage over wide open areas and migrate long distances. Conversely, most vespertilionids have low aspect ratios, indicative of broad wings useful for slow, "fluttery" flight. These bats typically forage along vegetation, forest canopies, narrow canyons, and over streams and ponds where slow and highly maneuverable flight is required. Exceptions to this generalization among the vespertilionids include the highly migratory species, such as the hoary bat and silver-haired bat.

ECHOLOCATION AND VOCALIZATIONS

Bats produce a wide variety of sounds designed for communication, some of which are well within our normal audible range, while others are ultrasonic and above the human range of hearing. Low frequency "squeaks and squawks," which can usually be heard by humans, are used to facilitate social interactions, such as territorial spacing among individuals, mother/infant communication, recognition, and warning calls. Bats use high frequency, ultrasonic sounds (generally above 20 kilohertz) outside our hearing range to navigate and avoid obstacles and capture prey in the dark. This capability of producing ultrasonic pulses and interpreting the echoes rebounding from objects in their path is called echolocation. Not all bats echolocate, however. Most megachiropterans cannot, although one genus (*Rousettus*) has developed a crude echolocation system. Echolocation is highly developed in the microchiropterans, where it is the primary means of orientation for most nocturnal, insectivorous species.

Echolocation calls originate in the larynx and may be emitted orally, as in the vespertilionids, or nasally, as in the phyllostomatids. Apparently a few species do produce echolocation pulses that can be heard by humans. For example, the spotted bat (*Euderma maculatum*) produces signals that sweep from 15 to 9 kilohertz, which is a faintly audible range for most people. The complex ultrasonic pulses are species specific and may incorporate a combination of frequencies in differing sequences, intensities, and duration to suit flying conditions. A bat traveling from its roost to a feeding area may produce 5 pulses each second, but the same individual trying to catch an insect may increase its rate of pulse production to more than 200 per second as it closes in on its prey. Such bursts of calls are known as "feeding buzzes."

Although the mechanisms of bat echolocation are not well understood, the efficacy of this method of "seeing" is evident in the great success of nocturnal bats. The maximum range of the echolocation "apparatus" is about 15 m, but within that distance bats are able to determine the direction, distance, velocity, shape, size, and even texture of their prey while tracking and closing in for the kill. To measure the distance to a target, it is believed that bats compute the time elapsed between sound production and echo return, basing the distance on the speed of sound. The details of the target (shape, texture, movement, etc.) are probably determined by comparing the frequencies of the original call with those of the echo. All of this information is processed in the auditory regions and associated structures of the bat's brain, undoubtedly involving an overwhelming array of neural operations. The bat likely broadens the area from which it collects information by moving its head.

In summarizing several studies of bat echolocation, Hill and Smith (1984) indicate that objects as small as 0.08 mm in diameter (approximately the diameter of human hair) are detected by especially sensitive species. The little brown myotis (*Myotis lucifugus*) is capable of capturing fruit flies (*Drosophila* spp.) at the rate of one every three seconds in a closed room, and one every seven seconds in the wild.

The unusual facial structures present on many bats are thought to be important in echolocation, although the exact function of these appendages is not clear. One example of such a structure is the nose leaf characteristic of phyllostomatid bats. This projection may serve to direct the nasally emitted calls, or it may help block the outbound soundwaves from the ear and thus increase sensitivity to reflected sound. The tragus may also serve either, or both, of these functions.

Several insects, most notably moths and lacewings, have ears sensitive to ultrasonic sounds; consequently their chances of being caught by bats may be up to 40 percent lower than those of "deaf" insects. The moth's ear provides information about the distance to the bat according to the bat's call. A faint call from a distant bat prompts the moth to turn and fly away. A more intense call from a closer bat causes the moth to fold its wings and dive toward the ground as an evasive maneuver. Some tiger moths have taken this anti-bat strategy one step further by having noisemakers on their chests which jam the bat's echolocation device.

VISION AND OLFACTION

A common misconception about bats is that they cannot see and must rely solely upon echolocation to orient themselves. To be "blind as a bat" is a poor analogy, however, as all bats have eyes and many can see quite well. The megachiropterans (which are diurnal in their habits and do not echolocate) have very good vision. Even those microchiropterans with highly developed echolocation systems have retained their eyes, although they are often tiny. These bats probably have vision adapted for low light conditions and may indeed have poor visual acuity, but nevertheless they can see. Some species, such as the pallid bat, have the ability to discriminate visual patterns, even in extremely low light. Vision may be important in recognizing landmarks among species that forage over long distances or that undertake extended seasonal migrations.

The role of olfaction, or the sense of smell, in bats is not clear. Many bats have large and highly developed glands that secrete odoriferous substances, and others have distinctive body odors. Olfaction is thought to play an important role in the courtship ritual of big-eared bats (genus *Plecotus*), two of which occur in Texas (Townsend's big-eared bat, *Plecotus townsendii*, and Rafinesque's big-eared bat, *Plecotus rafinesquii*). Both of these species possess pairs of large glands protruding from the rostrum. Male *P. townsendii* have been seen to rub the snout vigorously over the female's body prior to mating (Pearson et al., 1952).

The Brazilian free-tailed bat (*Tadarida brasiliensis*), which is one of Texas' most ubiquitous bats, has such a distinctive body odor that roosts can often be located by smell. Whether this scent is useful in guiding bats to the roost is not known, but the odor is unmistakable to anyone familiar with it.

ROOSTS

When not foraging, bats rest, groom, and interact with other bats at sites known as roosts. Roosting sites may vary greatly according to species, season, and even time of day. Many species tolerate only very narrow environmental gradients of factors such as temperature, relative humidity, or light intensity, and will occupy only sites where such factors are stable within specific limits. Thus, certain bats may roost only in caves and rock crevices, others in hollow trees or foliage, while still others are more general in their living requirements and can be found at a variety of sites. Moreover, roosts used during the daylight hours are quite often different from sites used at night.

Bats may occupy roosts either singly or in small groups (such as tree bats of the genus *Lasiurus*), or tremendous numbers may inhabit a single site, as in the famous central Texas colonies of Brazilian free-tailed bats (*Tadarida brasiliensis*), which may contain up to 20 million individuals during summer population peaks. It is not unusual for several species to use the same roost site, but in these cases the bats are generally segregated by species, with different species using particular areas of the site.

Many bats raise their young in special roosts known as nursery roosts. The populous summer colonies of *T. brasiliensis*, which inhabit caves, are classic examples of nursery roosts. These bats arrive in spring and give birth to their offspring in the caves, where the young are subsequently placed in clusters on the ceiling. Clusters of over a million baby bats have been observed at several of these caves. After the young have matured in fall, the nursery roosts are abandoned. Adult males are rarely found in nursery roosts.

Bats that hibernate have very strict requirements for hibernacula. Seemingly small fluctuations in temperature or relative humidity may critically affect survival over long winter months, and changes in these factors will cause torpid bats to arouse and move to more favorable locations. Generally, temperatures must not drop below freezing and preferred sites seem to have high relative humidity. Although caves and rock crevices are the most common hibernacula, a few species hibernate in relatively unprotected sites like hollow trees and behind tree bark, notably the hardy eastern red bat (*Lasiurus borealis*). Buildings are also used as hibernacula. In eastern Texas, the Brazilian free-tailed bat hibernates almost exclusively in buildings.

Bats are extremely sensitive to disturbance at roosts and will quickly abandon a site if harassed. Thus, continued disturbances at natural roosting sites is to be discouraged. Bats may occasionally take up residence in buildings, but they can be evicted by locating and covering their entranceways. This method is generally simpler, cleaner, and less costly than killing the animals or using noxious chemicals.

THERMOREGULATION

Due to their small body size, bats have a large relative surface area from which energy (heat) readily escapes, and are therefore extremely sensitive to changes in their environment. They typically respond to the changing seasons by moving to more favorable areas (migration) or by becoming torpid to conserve energy (hibernation).

During hibernation, bats conserve energy by reducing their metabolism to the absolute minimum required to sustain life. The bat lives off body fat stored during the preceding autumn and awakens at periodic intervals to assess conditions of the hibernaculum. If conditions have become unfavorable the bat will move either to a different location in the same roost or to a completely different site. Such periods of arousal consume a great deal of energy and repeated arousals will endanger the bat's life as it quickly expends its stored fat reserves. Thus, exceptionally prolonged winters or disturbances at hibernacula may cause starvation and contribute to high mortality at winter roosts.

Many species of Texas bats become torpid during winter. The cave myotis (*Myotis velifer*), Townsend's big-eared bat (*Plecotus townsendii*), and pallid bat (*Antrozous pallidus*) are examples of species which hibernate in tightly packed clusters in caves. Clustering behavior is common in bats but not well understood. This behavior probably helps bats stabilize their body temperature against fluctuations in ambient conditions. Bats near the center of the cluster are best protected against changes, and probably are better able to conserve their precious supply of body fat.

Bats not able to hibernate often migrate to warmer latitudes at the onset of winter. Migration is not well studied in bats, and destinations, distances traveled, and navigation mechanisms are very poorly known. Many Texas bats are seasonal residents of the state, migrating to Mexico for winter. The best known of these is the Brazilian free-tailed bat (*Tadarida brasiliensis mexicana*), which arrives abruptly in spring, completes its cycle of raising young, and then departs equally abruptly in fall. These bats winter in Mexico, and migratory moves of up to 1,300 km have been recorded (Villa-R. and Cockrum, 1962). Most species of *Myotis* from Trans-Pecos Texas are apparently only summer residents of Texas. Other migratory species that occur in Texas include the hoary bat (*Lasiurus cinereus*) and the silver-haired bat (*Lasionycteris noctivagans*), which both move as far northward as Canada in the summer months and then southward to the southern United States and Mexico in winter. These bats have been recorded throughout the year in Texas during both the northward and southward phases of their migratory cycles.

REPRODUCTION AND LIFE EXPECTANCY

The majority of hibernating bats do not follow the basic mammalian reproductive pattern of fertilizing the eggs by sperm immediately after copulation. Instead they exhibit a reproductive pattern characterized by two phenomena—delayed ovulation (egg releases) and over-winter storage of sperm in the female reproductive tract. For these species most matings occur in the fall prior to hibernation, but fertilization does not take place then because eggs have not yet been released into the female's reproductive tract. Instead, sperm are stored in the female's uterus, where they remain alive throughout her winter dormancy. Eggs are shed and fertilized soon after females emerge from hibernation in the spring.

With delayed ovulation and fertilization, there is no waiting for males to produce sperm in early spring. This reproductive strategy allows maximum time for young to mature, to learn to fly, and to store fat for the upcoming winter. In addition, it permits adults to attain breeding condition and to copulate during late summer and fall when their energy supply (insects) is plentiful. Delayed ovulation and fertilization are widespread among Texas bats, including those of the genera *Myotis*, *Pipistrellus*, *Eptesicus*, *Nycticeius*, *Lasiurus*, *Plecotus*, and *Antrozous*.

In contrast to sperm storage in the female genital tract and delayed fertilization, at least one other Texas bat stores sperm in the epididymis of the male reproductive tract until copulation in the spring. The Mexican free-tailed bat (*Tadarida brasiliensis*) produces sperm in the late autumn and winter months, but does not copulate until early spring.

Bats give birth in a manner similar to that of most other mammals. The annual litter is usually produced in May or June, although parturition may occur from April through July, varying with species and climate. Litter sizes range from one to five young per pregnancy, but most species give birth to only one young per year. Among Texas bats, twinning is common in the southeastern myotis, western pipistrelle, pallid bat, silver-haired bat, hoary bat, and Seminole bat. The red bat and the two species of yellow bats usually have more than two young per litter, sometimes as many as five. All Texas bats produce only one litter annually, and a few species (for example, all three phyllostomatid bats) do not reproduce while living here.

Most bats assemble in nursery colonies, ranging in size from a dozen or so to several million bats, to bear and raise their young. Pregnant females typically locate in buildings, caves, mines, or other dark, secluded places for this purpose. No nest is built. The mother holds her head upward as the young is born, and the baby is received in a pocket formed by the interfemoral membrane. In those species with a poorly developed interfemoral membrane the female hangs her head downward when giving birth. Baby bats are large and well developed at birth. As soon as they are born they crawl to the mother's breast and attach to a nipple. Mother bats, like all mammals, produce milk in mammary glands to nourish young during their initial growth period. Females have only one pair of functional mammary glands.

Typically the young remain attached to the mother throughout the day but are left behind, usually in clusters, when the mother emerges to feed in the evening. Mothers return periodically throughout the night to nurse the small babies, but as the young mature the mothers return less frequently. Most bats apparently can recognize their offspring and pick them out from the clusters to nurse. Young bats grow rapidly and most are able to fly within a few weeks. As they become increasingly adept at flight and obtaining their own food, they depend less on their mothers. The maternity colonies begin to disperse as the young are weaned.

Bats partly compensate for their low reproductive rate by having a long life expectancy. A conservative estimate for the average life span of a bat after surviving its first year is about ten years. Records of fifteen-year-old and older individuals living under natural conditions are not uncommon. One bat has been known to survive at least thirty-two years in the wild. Such a reproductive strategy, involving a low reproductive rate and long life expectancy, is unusual for a small mammal. Similar-sized rodents, by contrast, typically have many young per litter and several litters each year, but they usually live no longer than a few years. Nevertheless, because of their much higher reproductive rates, rodents still end up producing more young per mother than bats.

POPULATIONS

Due to the nocturnal and secretive habits of bats, study of their population dynamics is difficult. Determining the age and sex structure of bat populations, or even obtaining accurate estimates of numbers, is complicated by many factors, including the location of roosts in unknown or inaccessible sites and the difficulty of catching and observing bats. Also, determining the age of bats is practically impossible beyond the simple classifications of "juvenile" and "adult." As a result, detailed information on population size, mortality, longevity, home range use, and migration patterns is sparse for most species.

Studies of banded bats have provided some information. Bat banding is similar to bird banding, although in bats the coded bands are usually placed on the forearm rather than around the leg. As marked individuals are eventually recovered they provide clues to the movements of the population. Several banding studies have been carried out with Texas bats (i.e. Davis et al., 1962; Short et al., 1960; Twente, 1955b). These studies have shown that Brazilian free-tailed bats (*T. brasiliensis*) from central Texas annually migrate to Mexico in winter and then back to Texas in spring. Twente (1955b) showed that cave myotises (*M. velifer*) hibernating in north Texas caves periodically move between caves in winter, and in summer do not always return to the same roost on succeeding nights.

In other parts of North America, band recoveries have shown that significant mortality occurs during the first year of life, after which bats may be very long-lived. Reports of individuals recovered twenty to thirty years after they were initially banded are common. As bats do not have many natural predators, adverse winter weather probably accounts for much of the toll on yearlings. No predator is known to prey exclusively on bats, although a variety of animals periodically attack and capture bats. These include many species of snakes, which often frequent caves, as well as falcons, hawks, owls, raccoons, opossums, and feral cats. In cave situations, bats that fall to the floor are often eaten by dermestid beetles.

Estimating population numbers is extremely difficult. For most species, no accurate estimates of numbers exist. In Texas, however, several species aggregate at particular caves and crude population estimates have been attempted. Davis et al. (1962) estimated population numbers of *T. brasiliensis* for central Texas caves by using a combination of counts inside the caves, trapping, and observing the density and duration of exodus flights. They estimated that between 95.8 and 103.8 million *T. brasiliensis* occupy Texas caves during summer, although they admitted their precision was low. Twente (1955a) estimated numbers of bats in north Texas and Oklahoma caves by counting individuals in clusters and measuring the sizes of clusters.

PARASITES

Bats may be affected by a wide variety of parasites, most of which are specific to bats and not contagious to humans. These include both endoparasites, such as protozoans and helminths, and ectoparasitic arthropods.

Protozoans are single-celled animals and are generally transmitted through the bites of intermediate hosts such as mosquitoes and flies. Protozoans responsible for such diseases as malaria, sleeping sickness, Chagas' disease, murrina, and Mal de Caderas have been found in bats from around the world; however, such pathogens isolated from bats do not appear to cause human infection (Hill and Smith, 1984). Helminths, or parasitic worms, are internal parasites with complex life cycles usually requiring mollusks (snails) as hosts for the larval stages. Tapeworms may be found in the intestinal tract, flukes in the liver, lungs, or gall bladder, and roundworms may occupy almost any region of the body.

Ectoparasitic arthropods that subsist on bats include mites, batflies, ticks, kissing bugs, and fleas. Many of the mites and batflies are host specific, meaning they are associated only with particular bat species. Bat mites may also occupy only specific sites on their host, such as the oral mites (*Radfordiella*) of the Mexican long-nosed bat (*Leptonycteris nivalis*), which occurs in Trans-Pecos Texas. These mites are found only between the teeth and gums of this bat. Other locations for bat mites include in or at the bases of the ears, at the base of the tail and around the genitalia, on the wing, and around the eyes.

Batflies (family Nycteribiidae and Streblidae) are dipterans that have become especially adapted for living on bats. Nycteribids have completely lost their wings and so are incapable of flight, whereas the streblids have retained wings (though these are much reduced) and are capable of short flights. Both of these families of flies feed upon the blood of bats and their larvae are deposited directly on the bat, rather than an intermediate host.

FOOD

As a group, bats have the most diverse food habits of any group of mammals. Although there are no truly herbivorous (plant-eating) bats, the Chiroptera includes members that specialize in almost every other diet conceivable. There are bats that feed exclusively upon insects and other arthropods (insectivory), flesh (carnivory), fish (piscivory), fruit and flowers (frugivory), pollen and nectar (nectarivory), and blood (sanguivory), as well as bats that eat a variety of food items (omnivory). The bats of Texas are predominantly insectivores (twenty-nine species), but two nectarivores and one sanguivore also occur here.

Probably the most familiar of these specialized food habits is that of the true vampires, which feed exclusively on the blood of other vertebrates. Worldwide, there are only three species of these unusual bats, and all are endemic to the New World tropics and subtropics. One of these species, the hairy-legged vampire

(*Diphylla ecaudata*), has been recorded in Texas, although only once. The incisor teeth of vampire bats are enlarged and blade-like and are used to inflict a small wound from which blood is lapped with the tongue. Vampires also have dramatically modified stomachs, which vary from a baglike to a narrow, tubular structure with thin, elastic walls that easily stretch to accommodate large quantities of blood.

Equally interesting are the two nectarivorous bats (*Leptonycteris nivalis* and *Choeronycteris mexicana*) known to occur in Texas. These bats, both of which are rare in our state, exhibit structural modifications that allow them to feed upon the nectar and pollen of flowering plants. Nectarivores show an overall reduction in dentition, elongation of the rostrum, an extremely long, protrusible tongue for extracting nectar; and these bats can hover in flight much like hummingbirds.

Insectivorous bats, which constitute the great majority of the Texas bat fauna, eat a wide variety of night-flying and ground-dwelling insects, including moths, beetles, grasshoppers, flies and mosquitoes, termites, lacewings, crickets and katydids, cicadas, true bugs, and night-flying ants. The size range of insects eaten by bats varies from small gnats, weighing no more than about one-fifth of a milligram, to large moths weighing up to 200 mg.

To capture insects, bats will swoop low over the surface of bodies of water, snap prey out of the air, and even land on the ground to pursue them "on foot." Although some insect-eating bats grab their victims directly in their mouths, others scoop prey out of the air with their wings or the membranes enclosing their tails. The red bat (*Lasiurus borealis*), for example, performs an aerobatic somersault while closing its wings and tail membranes around the captured insect. Before resuming flight, it tucks its head down into the pocket formed by the tail membrane to collect the insect in its mouth (Hill and Smith, 1984).

Variety truly is the spice of life for insectivorous bats, whose menu may change from night to night and season to season. There is little evidence that any species of bat specializes on a particular type of insect, but some species seem to show distinct preferences for certain groups of insects. For example, Black (1974) investigated an insectivorous bat community in New Mexico and found that moths and beetles were the two insect groups most frequently used by the sixteen species of bats on his study site. By using a percent frequency occurrence of 65 percent or greater as a criterion for differentiating food habits, Black was able to classify most species as either "moth strategists" or "beetle strategists."

Insect-eating bats use their long canines to seize and pierce the prey, which is then reduced to minute fragments by the sharp-edged premolars and blade-like crests of the molar teeth. The sharp cusps and ridges of the opposing teeth act as scissors to cut up the insect food into tiny pieces. Small, soft insects are probably chewed and swallowed directly. Larger insects, especially beetles with hard, tough exoskeletons, require some processing before they can be consumed. Some large bats will carry their victims to a convenient night roost to be devoured.

Insectivorous bats, being primarily nocturnal in their foraging habits, become active at varying intervals after sunset. Smaller bats, such as the pipistrelles, generally emerge early in the twilight period, while larger bats, such as the mastiffs, tend to be the last to commence foraging. Smaller crepuscular bats also appear to fill their stomachs more quickly and return from foraging earlier than do larger species.

Many insectivorous bats seem to be rhythmic in their feeding, having two major feeding periods. The first feeding period begins with the onset of twilight, or shortly thereafter. Apparently the more efficient feeders in a given species are satiated by 10:00 P.M. while the feeding period appears to continue until about 12:00 P.M. for others less efficient. Then there is a period of relative inactivity until close to 3:00 A.M. However, some activity by stragglers may continue into the early daylight hours.

Three basic prey capture methods have been described for bats (Ross, 1967). Vision is important for those species, such as the pallid bat (*Antrozous pallidus*), which rely on ground- or vegetation-dwelling prey. Others, such as the western pipistrelle (*Pipistrellus hesperus*) and the Brazilian free-tailed bat (*Tadarida brasiliensis*), are "filter feeders" that appear to seize prey at random within dense flights of insects. Finally, some bats, such as the hoary bat (*Lasiurus cinereus*) and Townsend's big-eared bat (*Plecotus townsendii*), use individual pursuit to catch prey. In this type of predation, the bat approaches the prey from the rear and engulfs its abdomen.

Bats consume large quantities of insects. It has been estimated that they eat from a quarter to half of their body weight in insects nightly (Hill and Smith, 1984). A 20-gram bat may thus eat 5–10 g of insects in one night. At this conservative feeding rate such a bat might consume 1.8–3.6 kg in one year.

PUBLIC HEALTH

Although bats are highly secretive and retiring creatures, they occasionally come into contact with people. Usually, this happens when a few bats take up residence in a building or a single one is found on the ground. Their unusual appearance and our fear of rabies inevitably lead to consternation in such events, but the stigma bats carry of being dangerous is unwarranted.

Bats themselves can be afflicted with a variety of diseases and parasites, most of these contagious only to other bats. Only two diseases, rabies and histoplasmosis, are known to have been transmitted by bats to people, and fears of acquiring even these from bats are often grossly exaggerated.

Rabies. Since 1984, mammalogists at Texas A&M University have identified bats reported to the Texas Department of Health (TDH). These data (table 7) have contributed to knowledge of the distribution of many species, as well as having furnished information on the prevalence of bat rabies in Texas. Of the seventeen species so far reported, only ten included individuals that tested positive for the rabies virus.

During the period February, 1984, through February, 1987, the eastern red bat (*Lasiurus borealis*) was reported more often (626 specimens) than any other species and only 46 (7%) of these tested positive for rabies. The hoary bat (*Lasiurus*

TABLE 7. Summary of bats reported to the Texas Department of Health, February, 1984, to February, 1987

Species	Records	Rabid	% Rabid
P. rafinesquii	2	1	50
L. cinereus	40	10	25
T. brasiliensis	430	105	24
P. subflavus	10	1	10
L. seminolus	186	17	9
L. intermedius	126	11	9
L. borealis	626	46	7
E. fuscus	19	1	5
N. humeralis	221	6	3
M. velifer	82	2	2
M. austroriparius	4	0	0
M. ciliolabrum	1	0	0
L. noctivagans	5	0	0
P. hesperus	1	0	0
L. ega	1	0	0
A. pallidus	2	0	0
N. macrotis	2	0	0
Total	1758	200	11

cinereus) had the highest incidence of rabies (25%), but only 40 were reported to the TDH during the three-year period. A larger sample of these bats would be required before the susceptibility of hoary bats to the rabies virus could be evaluated objectively.

The most interesting data were those for the Brazilian free-tailed bat (*Tadarida brasiliensis*), which was the second most prevalent bat reported to the TDH. Of 430 *T. brasiliensis* reported, 105 (24%) tested positive for the rabies virus. Thus, this bat is more commonly encountered by people than most other species, and it also seems to be susceptible to rabies. Statewide, the summer population of *T. brasiliensis* is estimated to be around 100 million bats, and from the small sample it is obvious that a minute proportion of these bats come into contact with people.

For reasons of sampling bias, it would be erroneous to conclude that 11 percent of the bat population in Texas is infected with rabies. The TDH tests only bats that are submitted because they are rabies suspect; typically, these are bats found on the ground or in unusual places. Thus, the sample tested is not representative of the bat population as a whole. It is probable that less than half of 1 percent of bats contract rabies, a frequency no higher than that seen in many other animals (Tuttle, 1988). Furthermore, it appears that rabid bats seldom transmit rabies to any animal except other bats, and that the infected animals soon become paralyzed and die.

Logically, only injured or ill bats are likely to be encountered by people. For this reason, bats should not be handled without leather work gloves. Children should be warned never to pick bats up. If reasonable caution is taken, the danger of being bitten by a bat is very slim and, consequently, the probability of contracting rabies is incalculably small. However, one should not take unnecessary chances with a fatal disease and, in the event of any animal bite, medical advice should be sought immediately.

Histoplasmosis. This is a fungus disease affecting the lungs, with symptoms similar to those of tuberculosis. The fungus and its spores are associated with bird and bat droppings. Human infection occurs through inhalation of airborne spores, with the severity of infection normally proportional to the amount of dust-laden spores inhaled. Symptoms normally include a cough and resemble those of influenza. Bird droppings, frequently those of poultry or pigeons, are the primary source of infection for people. But the fungus may occasionally be found in droppings associated with hot, dry attics where bats roost. I am also personally familiar with a situation in which several biologists became infected with this disease after entering a cave housing a large nursery colony of Mexican free-tailed bats.

To avoid problems, one should be careful not to stir up and breathe dust in areas where bird or bat droppings have accumulated. When removing droppings or walking over a dry cave floor laden with droppings, use of a properly fitted respirator, capable of filtering particles as small as two microns in diameter, will greatly reduce the probability of exposure.

CONSERVATION

Drastic reductions in bat populations have occurred during recent years, not only in the United States, but worldwide. Several species of bats already are extinct, and others are near extinction. As a group, bats are exceptionally vulnerable to extinction because they typically rear only one young per year and, hence, are slow to recover from major population declines. In addition, many species form large aggregations vulnerable to mass destruction.

The causes for declining bat populations are not always certain and many possible factors may be responsible. However, there is little doubt that human beings have had a major adverse impact on many bat species by destroying habitat, outright extermination, vandalism, excessive disturbance at roosts and maternity colonies, pesticide poisoning, and use of other chemical toxicants.

Bats are highly sensitive to some insecticides, especially the chlorinated hydrocarbons. Bats feeding on insects treated with these chemicals slowly accumulate pesticide residues in their fat during late summer and fall, which eventually results in their death as this fat is broken down and used during hibernation and migration (Clark, 1981). To date only outright mortality on a local population level has been identified as a threat to bats from pesticides, but subtle—and equally devastating—effects are possible on such aspects as reproduction, acoustic behavior, and hibernation metabolism.

Human disturbance to hibernating and maternity colonies is apparently a major factor in the decline of many species. Increased numbers of visitors to caves appear to have a strong deleterious effect, since many bats are sensitive to intrusion and will leave a favored site after repeated disturbance. Frequently, well-meaning individuals, such as spelunkers (cave explorers) and biologists, are guilty of these disturbances. Unnecessary arousal of hibernating bats due to the presence of people may cause bats to awaken and use up precious fat needed to get them through the winter. Depleted energy supplies may cause many bats to starve before spring when their insect foods are again available.

Disturbance to summer maternity colonies can also be extremely detrimental. Maternity colonies are very intolerant of disturbance, especially when flightless newborn young are present. Baby bats may be dropped to their deaths or abandoned by panicked parents if disturbance occurs during this period.

In addition to the unintentional killing of bats, large numbers are deliberately killed and thousands of roosts are destroyed each year. Public health is the most common reason given for this destruction, even though bats rarely transmit diseases to humans. The removal of bats from the wild for scientific and educational purposes, as well as for food, also has been a contributing factor in declines of certain populations.

In recognition of the seriousness of bat population declines, many countries now offer bats some legal protection. In the United States six species are protected under the federal Endangered Species Act, and for one of these species, the Mexican long-nosed bat (*Leptonycteris nivalis*), the only known colony in the United States occurs in Big Bend National Park in Brewster County, Texas. In addition to the federal listing, the Texas Parks and Wildlife Department's list of protected non-game wildlife includes three species of bats (southern yellow bat, *Lasiurus ega*; spotted bat, *Euderma maculatum*; and Rafinesque's big-eared bat, *Plecotus rafinesquii*).

Some of the earliest efforts to give bats legal protection started in Texas. In 1917, the 35th Texas Legislature passed a general law (H.B. No. 40) making it a misdemeanor to kill or injure bats because of their perceived value in controlling malarial mosquitoes. The entire contents of that law, as adopted, were stated as follows (*General Laws of the State of Texas*, 35th Legislature, 1917, p. 124):

PROTECTION OF BATS

H.B. No. 40] Chapter 65

An Act making it a misdemeanor to kill or in any manner injure the winged quadruped known as the common bat; repealing all laws in conflict therewith, and declaring an emergency.

Be it enacted by the Legislature of the State of Texas:

Section 1. Article 887a. If any person shall willfully kill or in any manner injure any winged quadruped known as the common bat, he shall be deemed guilty of a misdemeanor and upon conviction shall be fined a sum of not less than five ($5.00) dollars nor more than fifteen ($15.00) dollars.

Sec. 2. All laws and parts of laws in conflict with the above provision shall be and the same are hereby repealed.

Sec. 3. The fact that there is now no law in this State protecting bats, creates an emergency and an imperative public necessity that the constitutional rule requiring bills to be read on three several days be suspended, and the same is hereby suspended, and that this Act take effect from and after its passage, and it is so enacted.

Approved March 9, 1917.

Takes effect 90 days after adjournment.

Effective conservation of bats will require the protection of hibernacula and maternity roosts and prevention of the general degradation of summer foraging habitats. A number of conservation efforts have been designed specifically to reduce disturbances to roosting bats and to increase available roosting sites. Special gates have been installed in front of cave entrances to allow easy passage for bats while excluding people. Interpretive signs have been placed at caves to warn against human disturbance to bat colonies. Bat houses and towers are being erected in forests and backyards to attract insect-eating bats, and caves located on private lands have been purchased to protect important bat colonies.

Efforts to construct artificial structures for attracting bats were also pioneered in Texas. Dr. Charles A. R. Campbell, a physician from San Antonio, was active for fifteen years or more in the establishment of artificial bat roosts because he believed bats were effective agents in the control of malarial mosquitoes. Dr. Campbell constructed several roosts near San Antonio, and their design was patented (U.S. Patent No. 1,083,318, issued January 6, 1914. *Patent Office Gazette*, Vol. 198, 1914, pp. 20–21).

In 1925, Dr. Campbell published a book describing his work with bats in which he urged communities to construct bat roosts (estimated to cost from $2,500 to $3,500) for controlling mosquitoes and harvesting guano. Campbell's book sparked considerable controversy. Professional mammalogists (Goldman, 1926) were skeptical about the accuracy of his statements and his conclusions. Tracy Storer visited the most successful of the bat roosts near Mitchell Lake, ten miles southwest of San Antonio, in 1919. Storer (1926) published a paper describing the roost, which had become inhabited by a colony of Mexican free-tailed bats (*Tadarida brasiliensis*). The roost was a tall, pyramid-shaped tower, measuring about twelve feet square at the base, about six feet square at the top, and about twenty feet in height (fig. 5). Storer obtained samples of guano from the roost and had them examined by several experts who found no evidence of any mosquito fragments. Some of the "bat towers" are still in existence today, and apparently they continue to attract bats (Merlin Tuttle, pers. comm.). However, Campbell's claims about their utility in malaria control and generating profits from the harvest of guano have never been substantiated.

Although conservation efforts such as those described above do provide some

Figure 5. Bat roost at Mitchell Lake, circa 1915–20, part of a malaria eradication program directed by Dr. Charles A. R. Campbell, who stands at the roost's foundation. A flatbed truck loaded with bags of guano is in the foreground. (Photo from *San Antonio Express-News*, courtesy University of Texas Institute of Texan Cultures, San Antonio)

protection, they will never be totally effective without changing people's attitudes toward bats. Negative attitudes must be replaced with an understanding of the ecological, economic, and scientific value of bats. This goal can be achieved only through education.

Concern about the worldwide decline of bats has led to the formation of Bat Conservation International (BCI), an organization whose purposes are to preserve bat populations around the world and to improve public attitudes toward bats. Having originally started BCI in Milwaukee, Wisconsin, Dr. Merlin Tuttle, its founder and science director, moved the organization to Austin, Texas, in 1986. According to Dr. Tuttle, the relocation of BCI was made, in part, because Texas harbors more species of bats than any other state in the United States and because the world's largest remaining bat cave (Bracken Cave) is located here. Information about this organization, including membership and subscription to its newsletter, may be obtained by writing to: Bat Conservation International, P.O. Box 162603, Austin, Texas 78716.

Several state and federal agencies and organizations are now actively involved in the bat conservation movement. These include the Texas Parks and Wildlife Department, Texas Natural Heritage Program, U.S. Fish and Wildlife Service, National Park Service, Texas Nature Conservancy, and the Texas Organization of

Endangered Species (TOES). TOES is a conservation organization devoted to the protection of rare and endangered plants and animals in Texas. As a private group of biologists, conservationists, and natural resource managers, TOES periodically publishes a watch-list of important species in need of conservation efforts. Their 1988 listing included five species of bats: the Mexican long-nosed bat (*Leptonycteris nivalis*), Rafinesque's big-eared bat (*Plecotus rafinesquii*), southern yellow bat (*Lasiurus ega*), spotted bat (*Euderma maculatum*), and southeastern myotis (*Myotis austroriparius*).

Scientific societies concerned with the conservation of Texas bats include the Texas Society of Mammalogists, Southwestern Association of Naturalists, American Society of Mammalogists, and the Texas Chapter of the Wildlife Society. All of these groups are working to reverse the bad image of bats and to promote these animals as an important part of our wildlife heritage.

Finally, people can help preserve our bat resource by following a few commonsense guidelines. Maternity colonies and hibernating bats should be avoided because even a slight disturbance could be harmful. If exploring a cave, try to leave everything as it was found and avoid disturbing or harming bats. Never shoot, poison, or otherwise harm bats. Bats are extremely beneficial insect predators, and nuisance bats generally can be encouraged to move elsewhere rather than killed. More and more people, when they discover that bats are beneficial and not dangerous, are going out of their way to attract bats by constructing backyard bat houses to take advantage of bats' insect-eating food habits. "Suggestions for Building Bat Houses and Attracting Bats" is a booklet available for one dollar from Bat Conservation International.

It is true that, given the opportunity, large numbers of bats may take up residence in attics or other parts of buildings and become a nuisance. In most cases, eviction and exclusion are the only safe, permanent remedies. When control is necessary, those measures which are most effective and least harmful to the bats should be selected. The most effective way to keep bats out of a building is to locate and seal the entrance through which they come and go. This should be done during the time of year when the bats are absent (usually October through March), or at night after the bats are out foraging. If blocking is not possible, lighting an attic can deter bats, but the light must be left on all the time. Poisons should rarely be used to solve bat problems in buildings, as they are usually unnecessary and may create far worse problems. A 1982 U.S. Fish and Wildlife Service publication (Resource Publication 143) entitled "House Bat Management," by Arthur M. Greenhall, is available and highly recommended as an aid in solving house bat problems.

2

Illustrated Keys to the Bats of Texas

Two keys have been prepared for the thirty-two species of bats known to occur in Texas. The first is based on external characters of adult animals. Subadult or juvenile bats usually show the diagnostic characters of the adult, although their pelage color may be slightly different (usually lighter). The best way to distinguish a juvenile from an adult bat is to examine the cartilaginous area in the finger (metacarpal-phalangeal) joints (fig. 6); the larger the area, signifying that the joints have not fused, the younger the bat.

After an identification is made using the external key, refer to the account of the species, which provides a more extended description including additional characters. The distribution maps also provide clues to help verify or question an identification. If a skull is available it should be examined as a check on the identification made by using external characters. The second key, based on cranial and dental structures, follows the key based on external characters.

Both keys are arranged so that there is always a choice between two statements (a couplet) about some characteristic of a bat. To identify a bat, use a millimeter ruler or similar instrument to make required measurements and select the appropriate alternative from each couplet (1, 2, 3, etc.) in the key. At the end of the statement chosen is a number that indicates the location of the next choice to be made. The process should be repeated until a name instead of a number is given at the end of a line.

KEY TO EXTERNAL CHARACTERS

A millimeter ruler and, in some cases, a hand lens are required. When possible, the characters used in this key have been chosen so that the animal need not be sacrificed, but best results are obtained when a scientifically prepared voucher specimen is available (see Barbour and Davis, 1969, for an explanation of this procedure).

All measurements (as defined and shown in figure 6) and weights are in millimeters and grams respectively, unless stated otherwise. One of the external measurements taken on bats, the length of the forearm, differs from the usual standard measurements taken on small mammals. Forearm length is the best indicator of size in bats as it can be measured with precision.

1. Distinct, upwardly and freely projecting, triangular-shaped nose leaf at end of elongated snout (fig. 7a) . . . Family Phyllostomatidae (in part) 2
   Nose leaf absent, indistinct, or modified as lateral ridges or low mound-like structure; snout normal (fig. 7b) 3
2. Tail evident, protruding about 10 mm from dorsal side of interfemoral membrane (fig. 8a); distance from eye to nose about twice distance from eye to ear (fig. 8c); forearm less than 48 mm .... *Choeronycteris mexicana*
   Tail not evident (fig. 8b); eye about midway between nose and ear (fig. 8d); forearm more than 48 mm .... *Leptonycteris nivalis*
3. Thumb longer than 10 mm; hair straight, lying smoothly, glossy tipped . . . Family Phyllostomatidae (subfamily Desmodontinae, vampire bats) .... *Diphylla ecaudata*
   Thumb less than 10 mm; hair slightly woolly, pelage lax, not usually lying smoothly, not glossy tipped 4
4. Prominent grooves and flaps on chin (fig. 9a); tail protruding from dorsal surface of interfemoral membrane (fig 9b) . . . Family Mormoopidae ... . *Mormoops megalophylla*
   No notable grooves or flaps on chin; lumps above nose or wrinkled lips possible, most faces lacking even these characteristics; tail extending to or beyond the edge of the interfemoral membrane 5
5. Tail extending conspicuously beyond free edge of interfemoral membrane (fig. 10a) . . . Family Molossidae 6
   Tail extending to the free edge of interfemoral membrane (fig. 10b) . . . Family Vespertilionidae 9
6. Forearm more than 70 mm; upper lips without deep vertical grooves (fig. 11a) .... *Eumops perotis*
   Forearm less than 70 mm; upper lips with deep vertical grooves (fig. 11b) 7
7. Forearm less than 52 mm 8
   Forearm more than 52 mm (58–64) .... *Nyctinomops macrotis*
8. Ears not united at base (fig. 12a); second phalanyx of fourth finger more than 5 mm (fig. 12c) .... *Tadarida brasiliensis*
   Ears joined at base (fig. 12b); second phalanyx of fourth finger less than 5 mm (fig. 12d) .... *Nyctinomops femorosacca*
9. Ears proportionally large, more than 25 mm from notch to tip 10
   Ears of normal size, less than 25 mm from notch to tip 13
10. Color black with three large white spots on back, one just behind each shoulder, the other at the base of tail (fig. 13) ..... *Euderma maculatum*
    Color variable, but not black; no white spots on back 11
11. Dorsal color pale yellow; no distinctive glands evident on either side of the nose .... *Antrozous pallidus*
    Dorsal color light brown to gray; distinctive glands (large bumps) evident on either side of the nose (fig. 14) 12
12. Hairs on belly with white tips; strong contrast in color between the basal portions and tips of hairs on both back and belly (fig. 15a); presence of long hairs projecting beyond the toes .... *Plecotus rafinesquii*
    Hairs on belly with pinkish buff tips; little contrast in color between basal portions and tips of hairs on back or belly (fig. 15b); absence of long hairs projecting beyond the toes .... *Plecotus townsendii*

13. At least the anterior half of the dorsal surface of the interfemoral membrane well furred (fig. 16) 14
    Dorsal surface of interfemoral membrane naked, scantily haired, or at most lightly furred on the anterior third 20
14. Color of hair black, with many of the hairs distinctly silver-tipped (fig. 4h) .... *Lasionycteris noctivagans*
    Color various, but never uniformly black 15
15. Color yellowish 16
    Color reddish, brownish, or grayish (not yellowish) 17
16. Total length more than 120 mm .... *Lasiurus intermedius*
    Total length less than 120 mm .... *Lasiurus ega*
17. Forearm more than 45 mm; color wood brown heavily frosted with white (fig. 4e) .... *Lasiurus cinereus*
    Forearm less than 45 mm; upper parts reddish or mahogany 18
18. Upper parts brick red to rusty red, frequently washed with white 19
    Upper parts mahogany brown washed with white ...... *Lasiurus seminolus*
19. Color reddish with frosted appearance resulting from white-tipped hairs; interfemoral membrane fully haired (fig. 4g) .... *Lasiurus borealis*
    Color rusty red to brownish without frosted appearance; posterior third of interfemoral membrane bare or only scantily haired .... *Lasiurus blossevillii*
20. Tragus (projection within ear) short, blunt, and curved (fig. 17a,b) 21
    Tragus long, pointed, and straight (fig. 17c) 23
21. Forearm more than 40 mm .... *Eptesicus fuscus*
    Forearm less than 40 mm 22
22. Forearm more than 32 mm; interfemoral membrane naked; color brown .... *Nycticeius humeralis*
    Forearm less than 32 mm; interfemoral membrane sparsely furred on anterior third of dorsal surface; color drab to smoke gray .... *Pipistrellus hesperus*
23. Dorsal fur tricolored when parted: black at base, wide band of light yellowish brown in middle, tipped with slightly darker contrasting color (fig. 15c); leading edge of wing membrane noticeably paler than rest of membrane .... *Pipistrellus subflavus*
    Dorsal fur bicolored or unicolored with no light band in the middle (fig. 15d); leading edge of wing same color as other parts of membrane 24
24. Calcar with well-marked keel (fig. 18a) 25
    Calcar without well-marked keel (fig. 18b) 27
25. Forearm more than 36 mm; foot more than 8 mm long; underside of wing furred to elbow (fig. 19); pelage dark brown .... *Myotis volans*
    Forearm less than 36 mm; foot less than 8 mm long; underside of wing not furred to elbow; pelage light brown to buff brown 26
26. Hairs on back with dull reddish brown tips; black mask not as noticeable; thumb less than 4 mm long; naked part of snout about as long as the width of the nostrils when viewed from above (fig. 20a) .... *Myotis californicus*
    Fur on back with long, glossy, brownish tips; black mask usually noticeable; thumb more than 4 mm long; naked part of snout approximately 1.5 times the width of the nostrils (fig. 20b) .... *Myotis ciliolabrum*
27. Forearm more than 40 mm 28
    Forearm usually less than 40 mm 29
28. Conspicuous fringe of stiff hairs on free edge of interfemoral membrane (fig. 21) .... *Myotis thysanodes*

No conspicuous fringe of stiff hairs on free edge of interfemoral membrane .... *Myotis velifer*

29. In Texas occurring west of 100th meridian (fig. 22) 30
 In Texas occurring east of 100th meridian (fig. 22) 31

30. Dorsal fur usually with a slight sheen; forearm more than 36 mm; total length more than 80 mm .... *Myotis lucifugus*
 Dorsal fur usually lacking a sheen; forearm less than 36 mm; total length less than 80 mm .... *Myotis yumanensis*

31. Ear more than 16 mm, extending more than 2 mm beyond nose when laid forward; tragus long (9–10 mm), thin and somewhat sickle-shaped .... *Myotis septentrionalis*
 Ear less than 16 mm, not extending more than 2 mm beyond nose when laid forward; tragus shorter and straight .... *Myotis austroriparius*

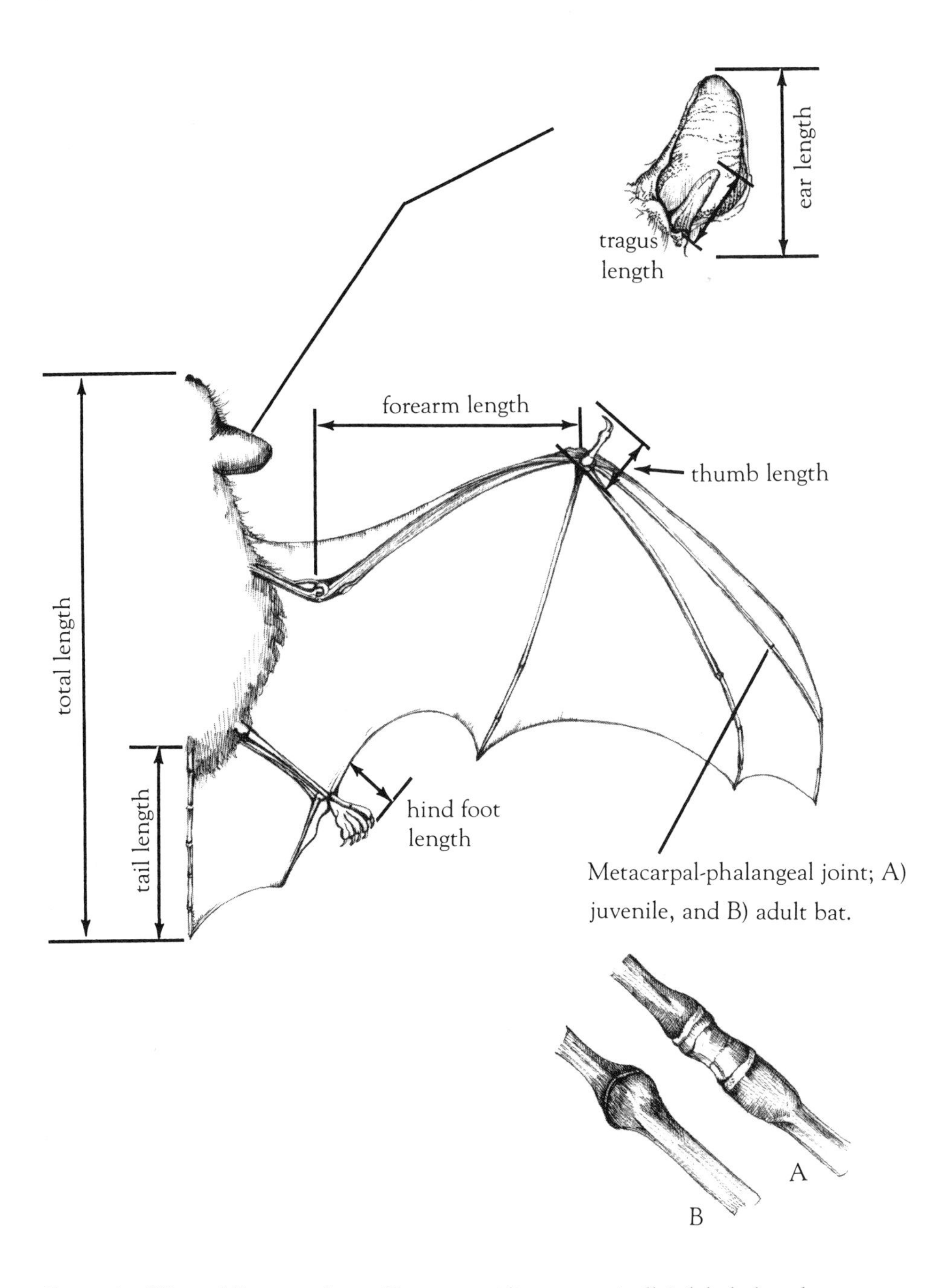

Figure 6. Wing of *Eptesicus fuscus* (drawn semidiagrammatically), labeled to show names of external parts and measurements used in key to Texas bats. The insert drawing is an enlargement of the metacarpal-phalangeal joint in a juvenile (A) and adult (B) bat.

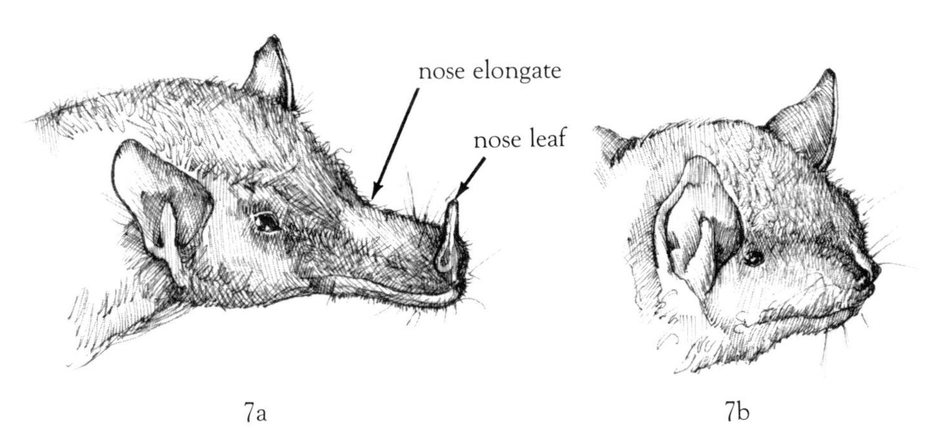

Figure 7.

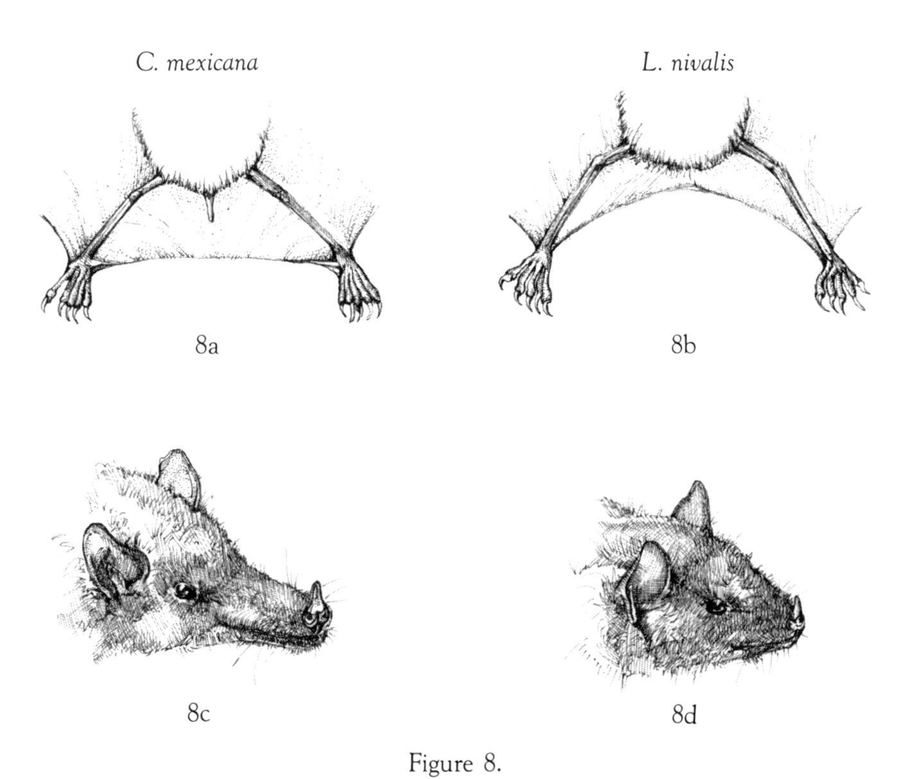

Figure 8.

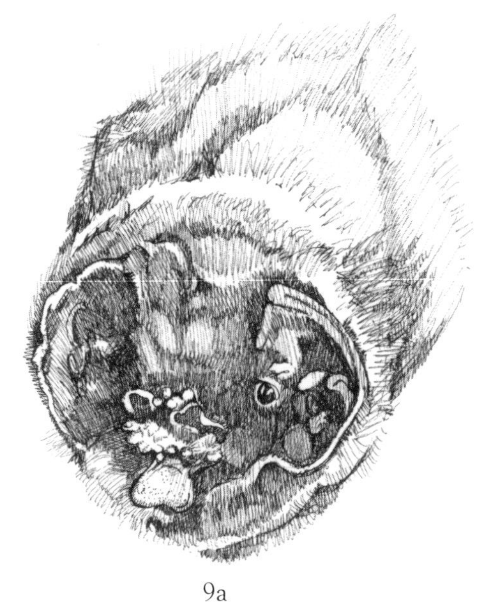

9a

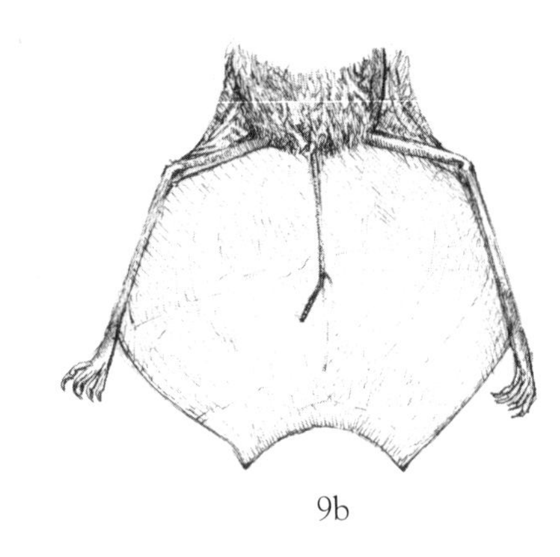

9b

Figure 9.

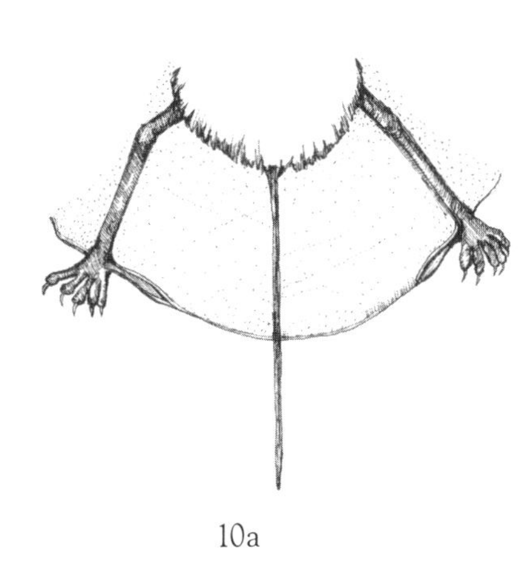

10a

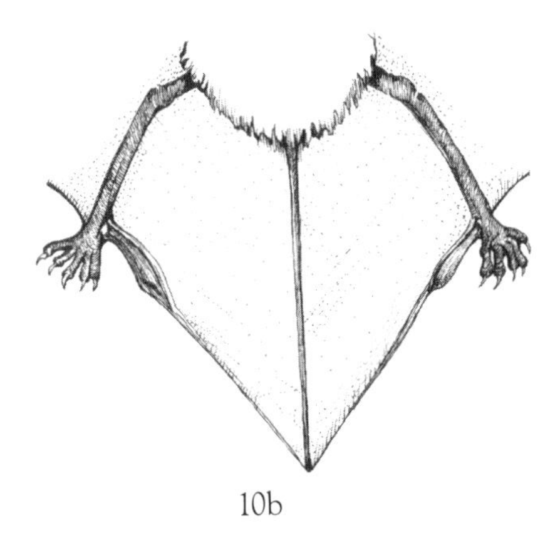

10b

Figure 10.

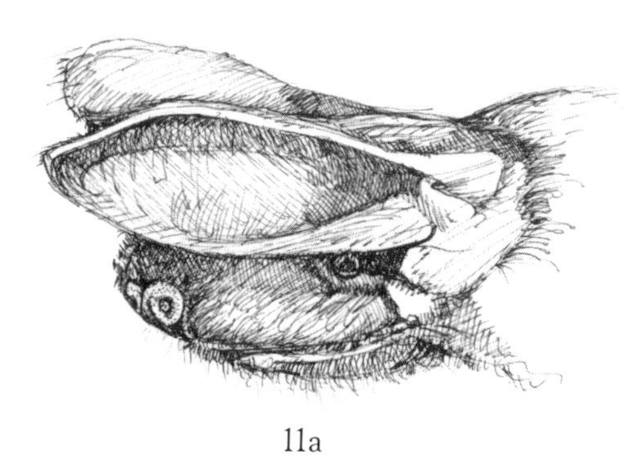

11a

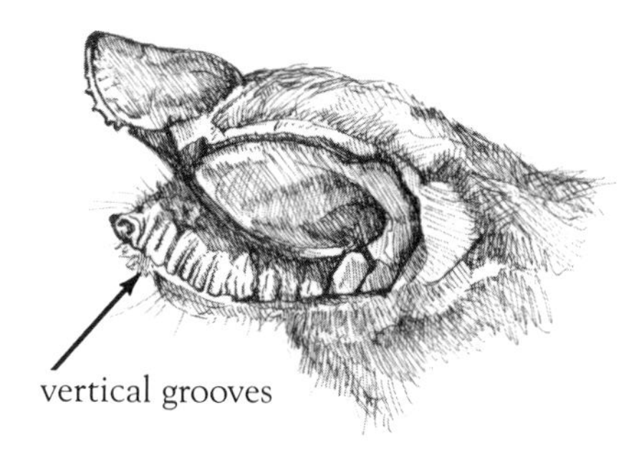

11b

Figure 11.

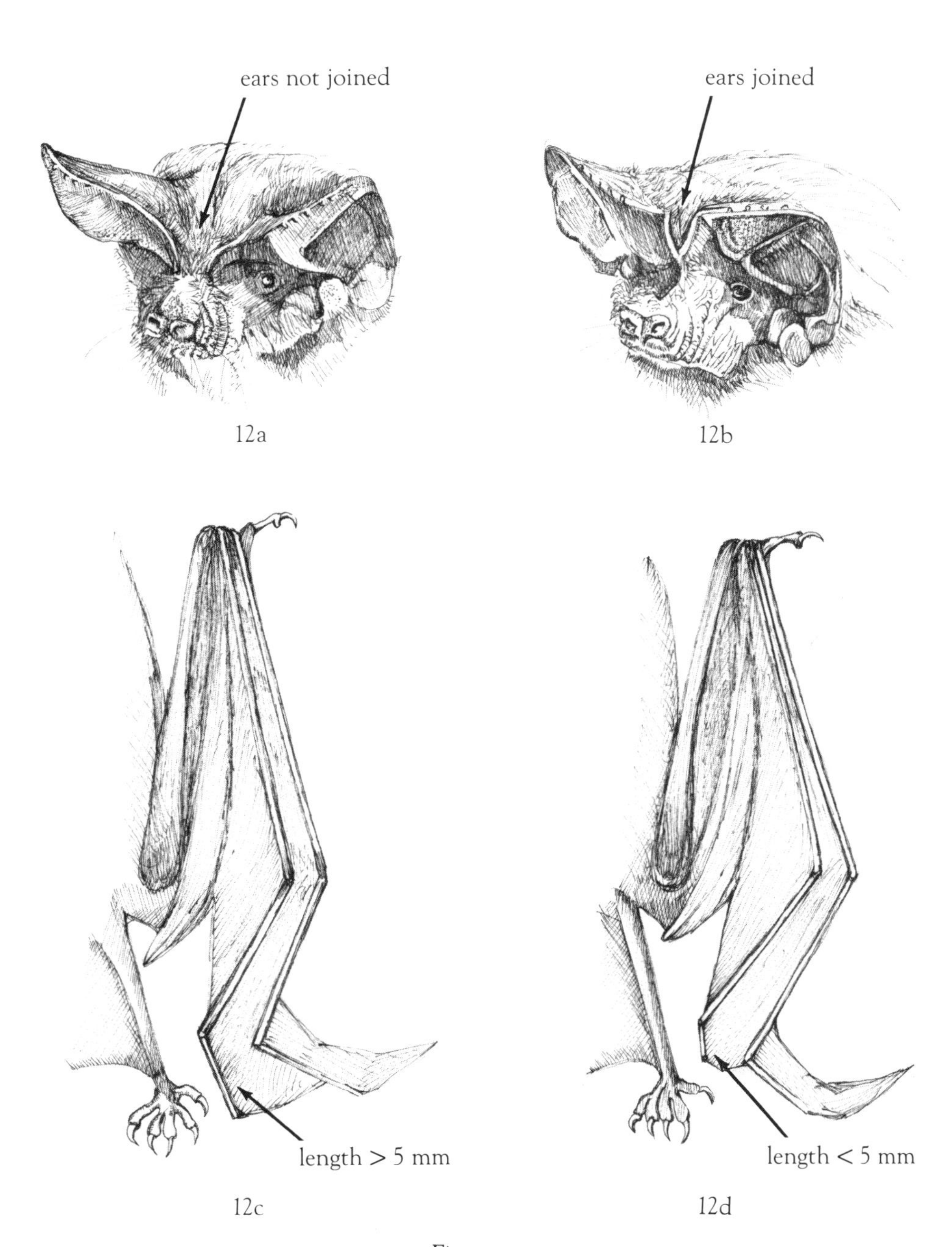
ears not joined
ears joined
12a
12b
length > 5 mm
length < 5 mm
12c
12d

Figure 12.

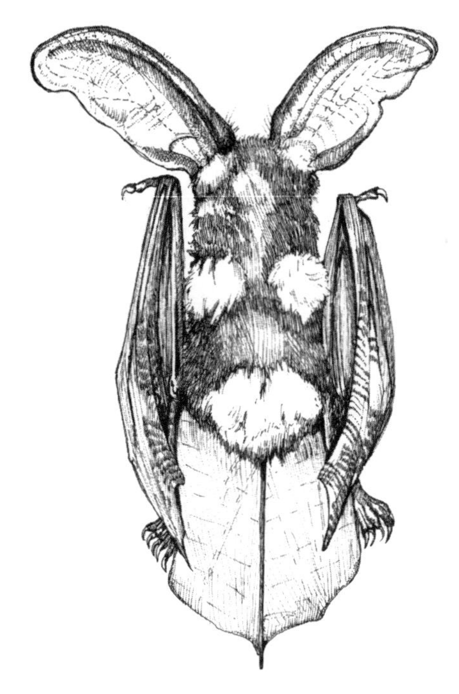

Figure 13.

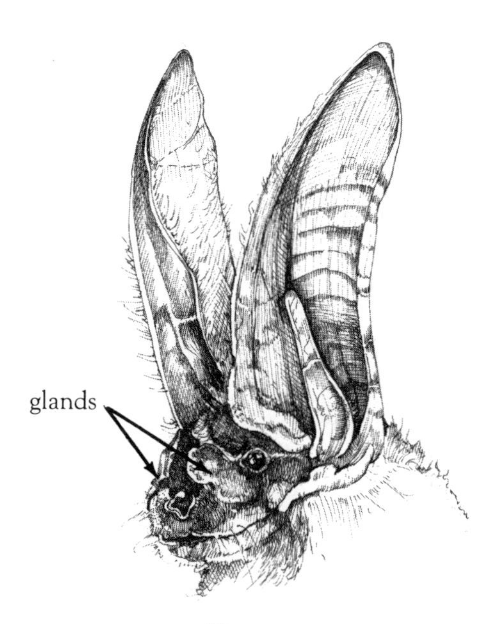

Figure 14.

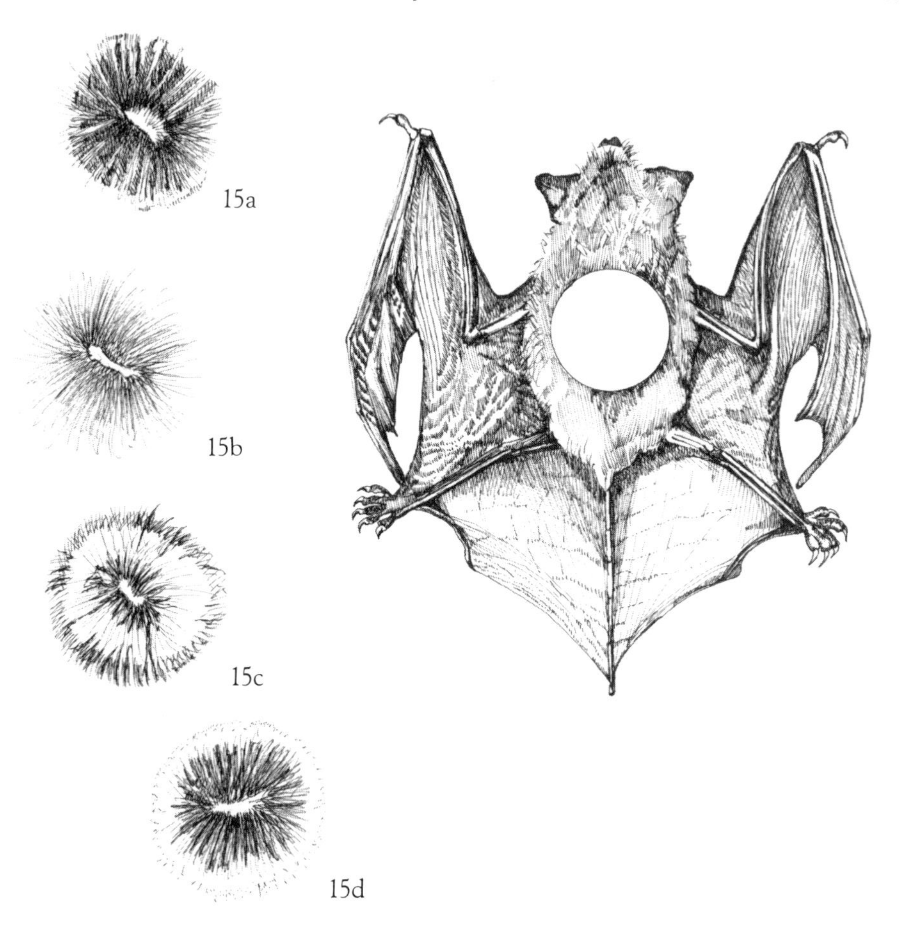

Figure 15.

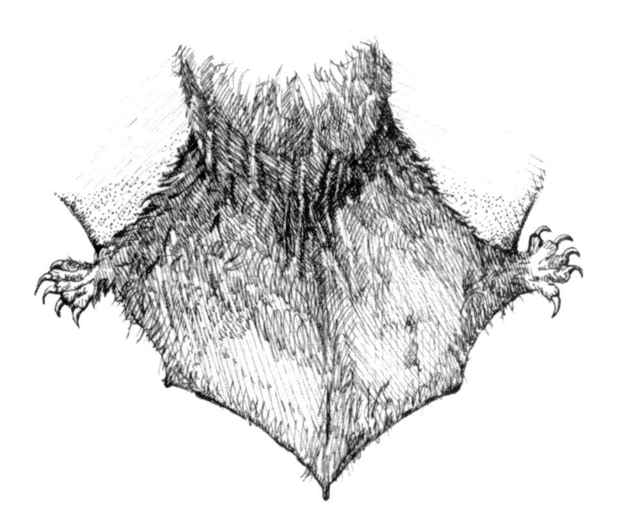

Figure 16.

a) broad, rounded tragus

b) curved, blunt tragus

c) straight, pointed tragus

Figure 17.

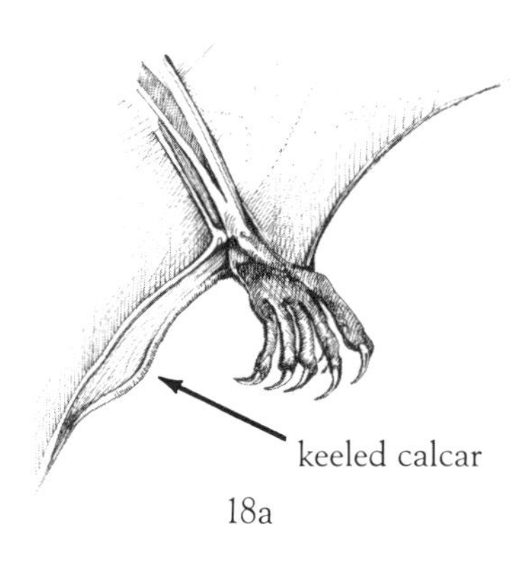

18a

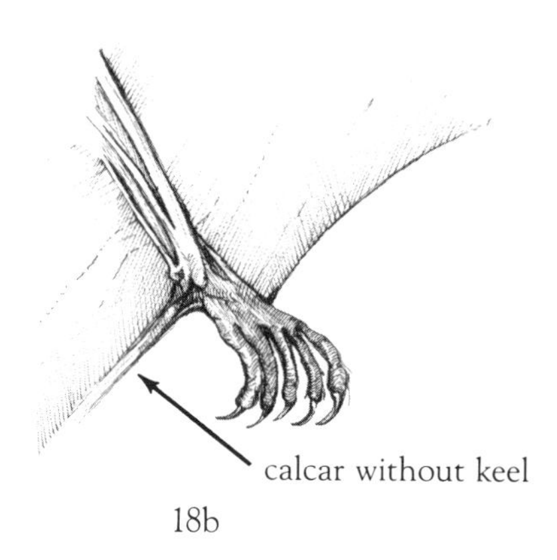

18b

Figure 18.

(Figure 19 opposite)

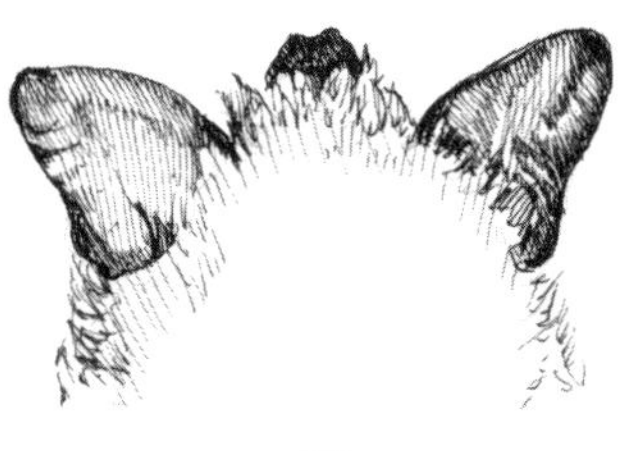

20a

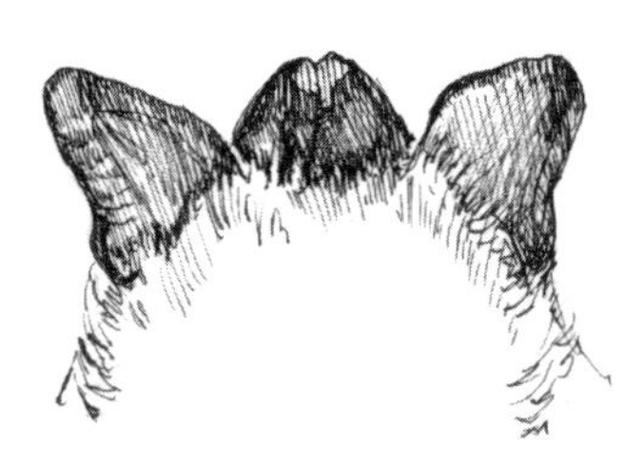

20b

Figure 20.

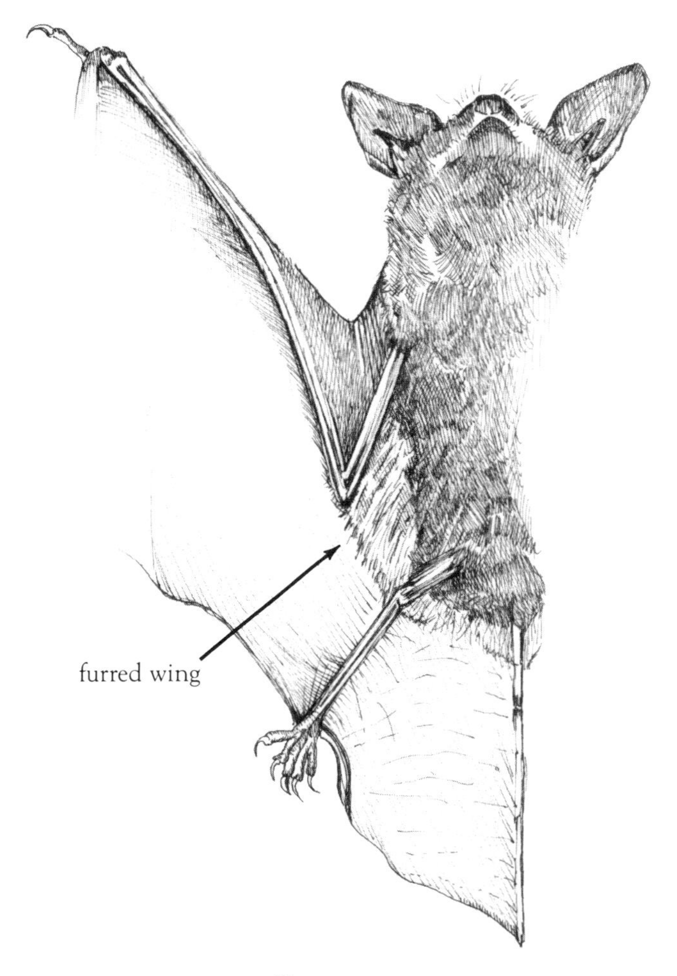

Figure 19.

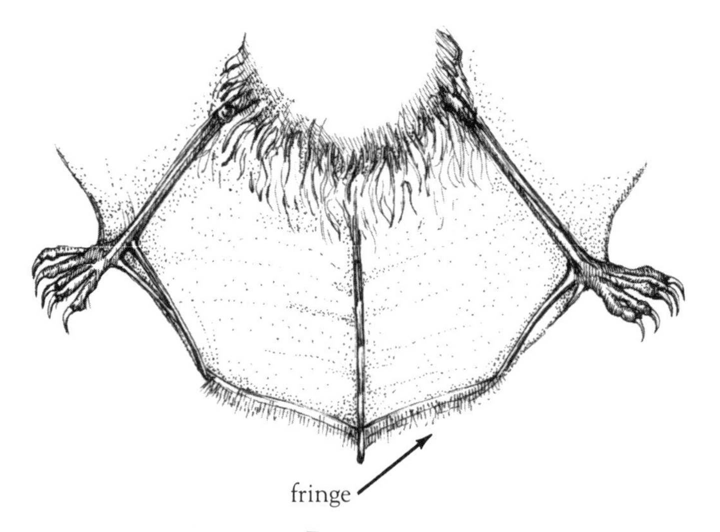

Figure 21.

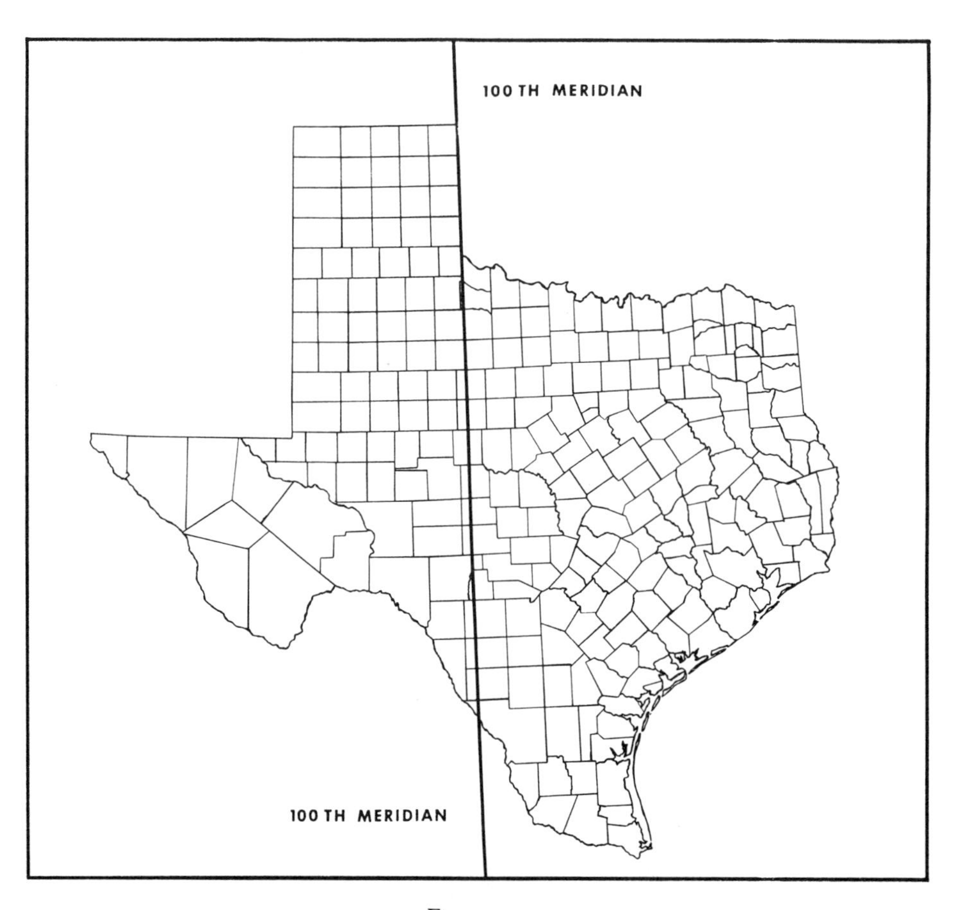

Figure 22.

KEY TO THE SKULLS

This key, which is based on cranial and dental features of adults, includes the same species that appear in the preceding key based on external features. The user will need a cleaned skull, a binocular dissecting microscope with ocular micrometer, and dial calipers capable of measuring to the nearest 0.1 mm. Measurements used in the key should be recorded as described in figure 23.

1. Upper incisors enlarged and blade-like; only three pairs of small teeth behind upper canines (fig. 24) .... *Diphylla ecaudata*
   Upper incisors small, much smaller than canines; more than three pairs of teeth behind upper canines (fig. 23) 2
2. Skull long and narrow; rostrum elongated, at least as long as braincase 3
   Skull not especially long and narrow; rostrum not greatly elongated 4
3. Rostrum exceptionally long; zygomatic arches of skull incomplete (fig. 25a); no lower incisors .... *Choeronycteris mexicana*
   Rostrum less elongate; zygomatic arches of skull complete (fig. 25b); two lower incisors .... *Leptonycteris nivalis*
4. Rostrum abruptly upturned giving a dorsal profile that is concave in lateral view (fig. 26a) .... *Mormoops megalophylla*
   Rostrum not abruptly upturned, dorsal profile of skull usually convex or flat (fig. 26b) 5
5. Six postcanine teeth in upper jaw (fig. 27a) 6
   Fewer than six postcanine teeth in upper jaw (fig. 27b–l) 14
6. Greatest length of skull more than 15.5 mm 7
   Greatest length of skull less than 15.5 mm 8
7. Rostrum relatively massive; sagittal crest well defined (fig. 28a); breadth across canines more than 4.3 mm .... *Myotis velifer*
   Rostrum relatively narrower and less massive; sagittal crest not well defined (fig. 28b); breadth across canines less than 4.3 mm .... *Myotis thysanodes*
8. Width across molars greater than 5.5 mm 9
   Width across molars less than 5.5 mm 12
9. Occurring east of 100th meridian (fig. 22) 10
   Occurring west of 100th meridian (fig. 22) 11
10. Interorbital breadth more than 4 mm .... *Myotis austroriparius*
    Interorbital breadth less than 4 mm .... *Myotis septentrionalis*
11. Slope from top of rostrum to top of braincase abrupt (fig. 29a) .... *Myotis volans*
    Slope from top of rostrum to top of braincase gradual (fig. 29b) .... *Myotis lucifugus*
12. Skull with forehead steeply sloping (fig. 30a, b) 13
    Skull with forehead gradually sloping (fig. 30c) .... *Myotis ciliolabrum*
13. Interorbital breadth more than 3.5 mm (fig. 23) .... *Myotis yumanensis*
    Interorbital breadth less than 3.5 mm (fig. 23) .... *Myotis californicus*
14. One pair of upper incisors (fig. 27f, h–l) 15
    Two pairs of upper incisors (fig. 27b–e, g) 26
15. One pair of upper premolars (fig. 27f, h, i) 16
    Two pairs of upper premolars (fig. 27f, j–l) 19
16. Two pairs of lower incisors (fig. 31h) .... *Antrozous pallidus*
    Three pairs of lower incisors (fig. 31f, i) 17

17\. Upper incisors in contact with canine (fig. 27f); third lower incisor crowded and smaller than first and second incisors 18
Upper incisors separated from canine (fig. 27i); third lower incisor not crowded, and equal in size to the first and second (fig. 31) .... *Nycticeius humeralis*

18\. Length of maxillary toothrow 6 mm or more .... *Lasiurus intermedius*
Length of maxillary toothrow less than 6 mm .... *Lasiurus ega*

19\. Breadth across canines greater than length of maxillary toothrow 20
Breadth across canines less than length of maxillary toothrow 23

20\. Greatest length of skull more than 15 mm .... *Lasiurus cinereus*
Greatest length of skull less than 15 mm 21

21\. Lacrimal ridge present (fig. 32a) 22
Lacrimal ridge not developed (fig. 32b) .... *Lasiurus seminolus*

22\. Greatest length of skull more than 12.8 mm in males and 13.0 in females; length of maxillary toothrow more than 4.2 mm in males and 4.6 in females .... *Lasiurus borealis*
Greatest length of skull less than 12.8 mm in males and 13.0 in females; length of maxillary toothrow less than 4.2 mm in males and 4.6 in females .... *Lasiurus blossevillii*

23\. Greatest length of skull less than 30 mm; premaxillary gap present on front part of bony palate (fig. 33a) 24
Greatest length of skull more than 30 mm; no premaxillary gap on front of bony palate (fig. 33b) .... *Eumops perotis*

24\. Greatest length of skull more than 21 mm .... *Nyctinomops macrotis*
Greatest length of skull less than 21 mm 25

25\. Greatest length of skull more than 18 mm .... *Nyctinomops femorosacca*
Greatest length of skull less than 18 mm .... *Tadarida brasiliensis*

26\. One pair of upper premolars (fig. 27g) .... *Eptesicus fuscus*
Two pairs of upper premolars (fig. 27b–e) 27

27\. Two lower premolars in each side of jaw (fig. 31c, d) 28
Three lower premolars in each side of jaw (fig. 31b, e) 30

28\. Auditory bullae very large and elongate; canines small and weak, the lower one with a distinct accessory cusp behind; zygomata abruptly widened in the middle (fig. 34a) .... *Euderma maculatum*
Auditory bullae neither greatly enlarged nor elongate; canines large and strong, unicuspidate; zygomata of uniform height throughout 29

29\. Skull nearly straight in dorsal profile; palate extending far behind molars fig. 34b) .... *Pipistrellus hesperus*
Skull concave in dorsal profile; palate extending little behind molars (fig. 34c) .... *Pipistrellus subflavus*

30\. Auditory bullae much enlarged; rostrum narrow, evenly convex above or slightly concave medially; forehead conspicuously elevated (fig. 35a,b) 31
Auditory bullae not especially enlarged; rostrum broad, concave on each side; forehead nearly flat (fig. 36a,b) .... *Lasionycteris noctivagans*

31\. First upper incisor with one cusp (fig. 37a) .... *Plecotus townsendii*
First upper incisor with two cusps (fig. 37b) .... *Plecotus rafinesquii*

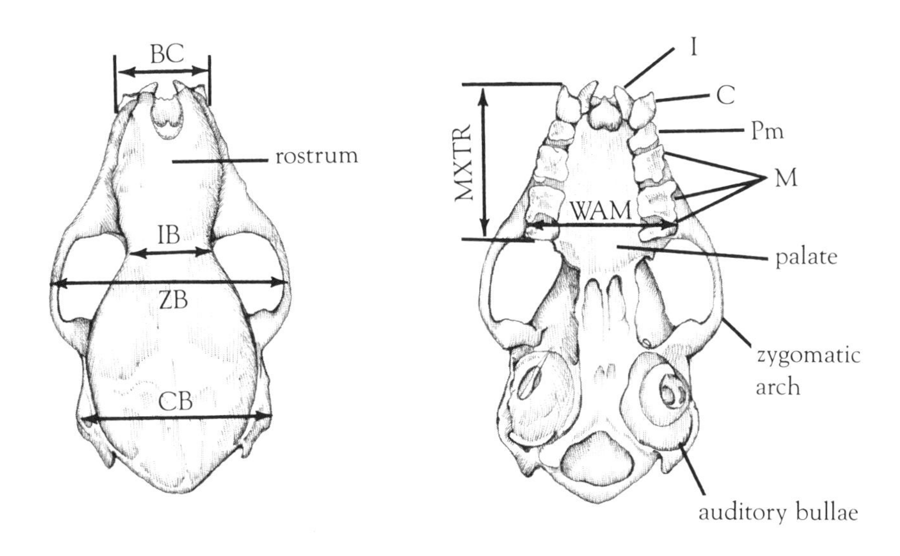

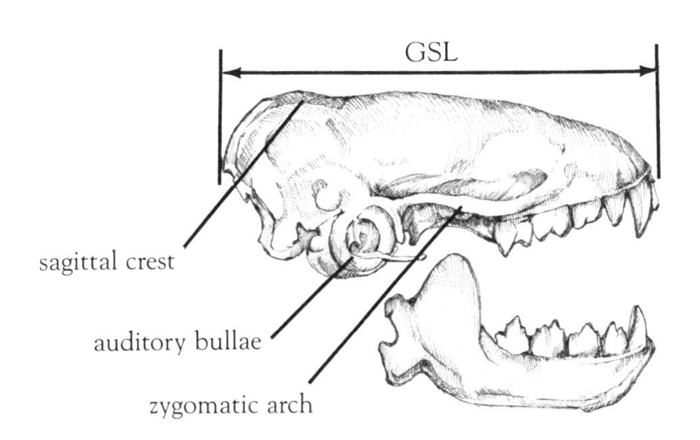

Figure 23. Skull of *Antrozous pallidus*, labeled to show the various parts and measurements used in the key to Texas bats. Skull measurements are as follows: GSL, greatest length of skull; BC, breadth across canines; IB, interorbital breadth; ZB, zygomatic breadth; CB, cranial breadth; MXTR, length of maxillary toothrow; WAM, width across molars. Teeth are labeled as follows: incisors, I; canines, C; premolars, Pm; and molars, M.

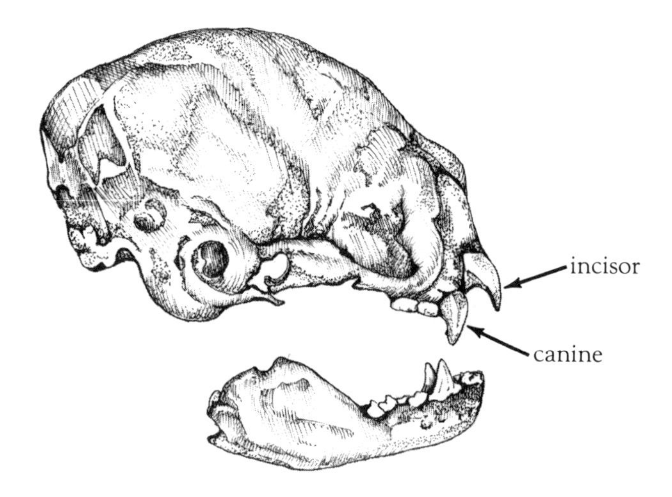

Figure 24.

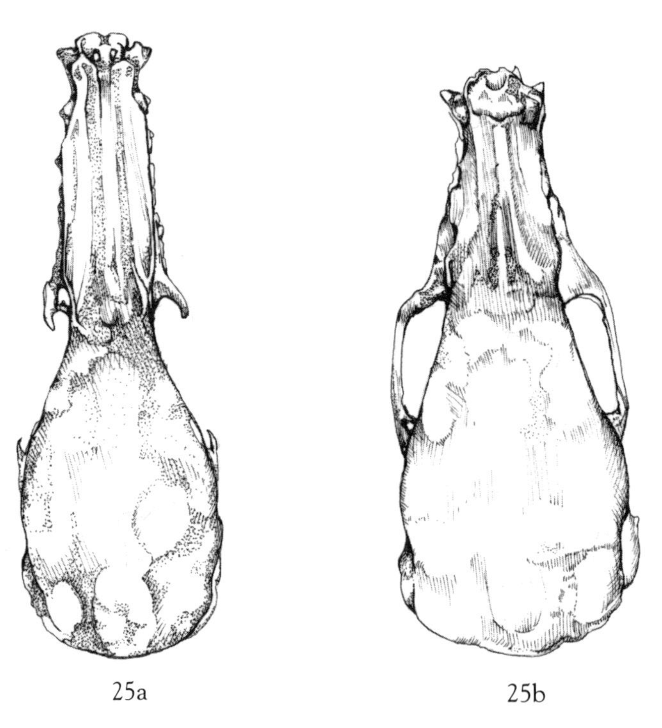

Figure 25.

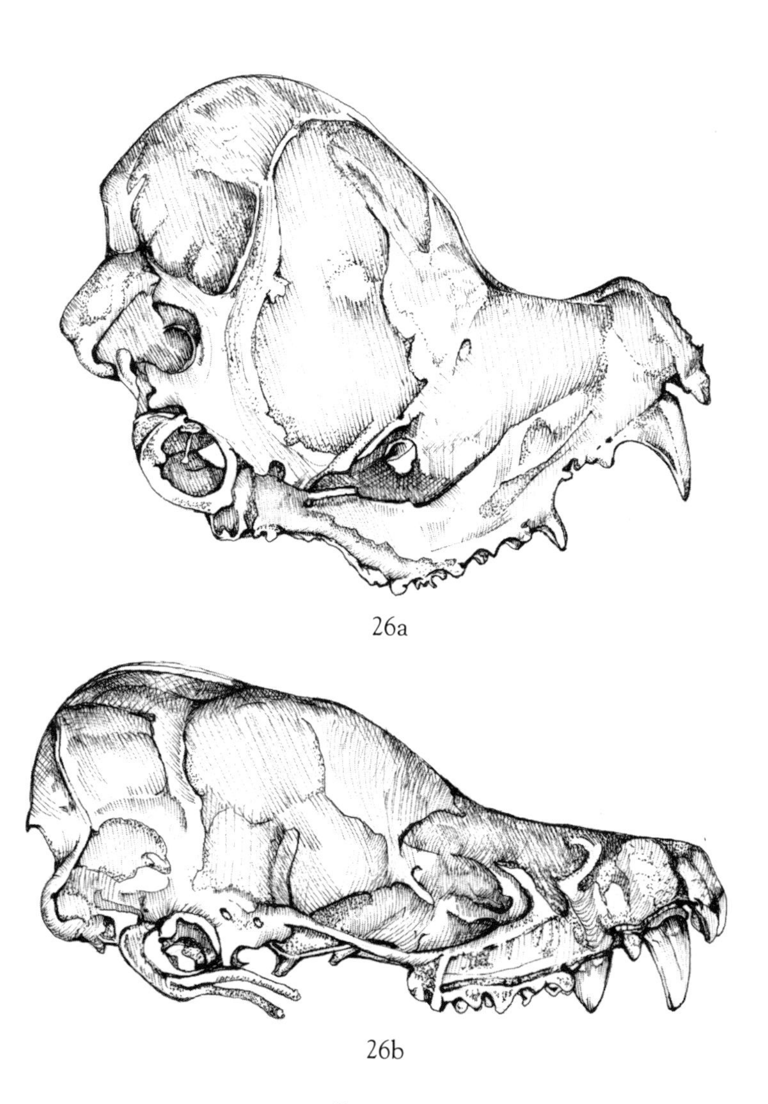

26a

26b

Figure 26.

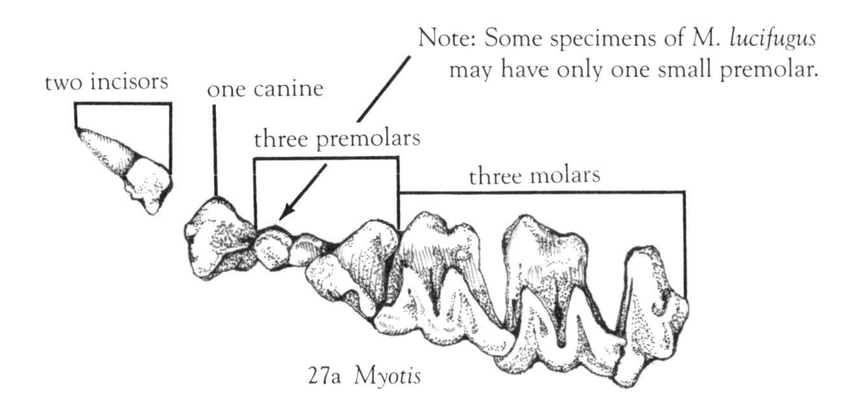

27a *Myotis*

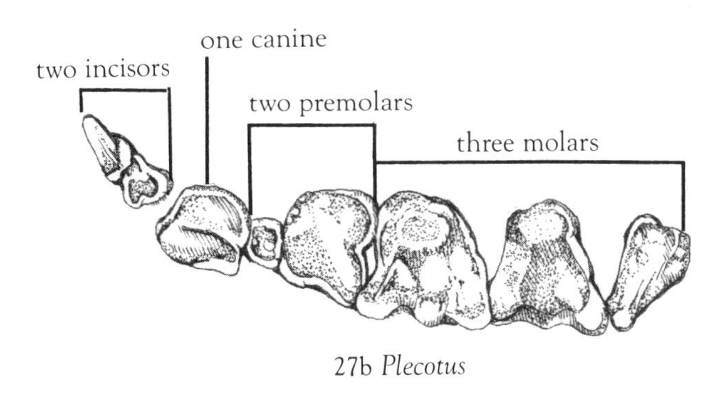

27b *Plecotus*

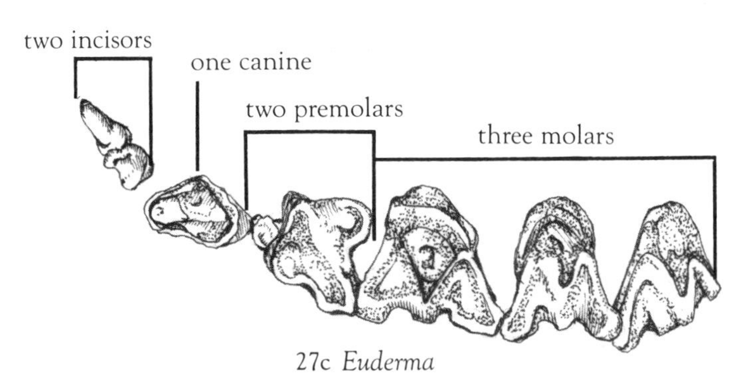

27c *Euderma*

Figure 27. Upper teeth of Texas bats

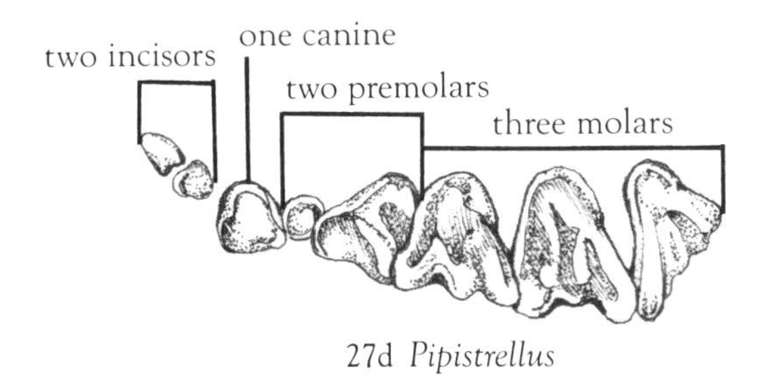

27d *Pipistrellus*

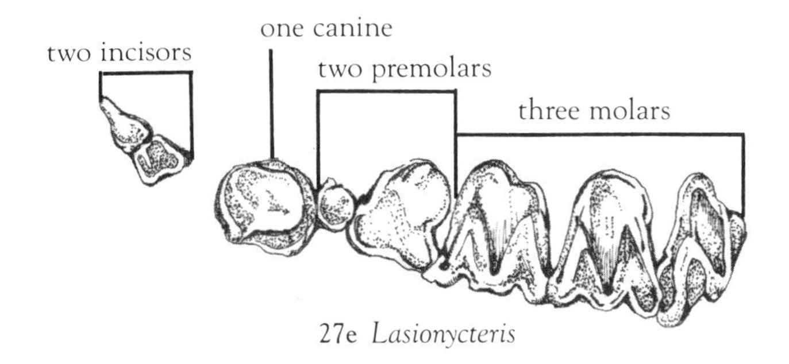

27e *Lasionycteris*

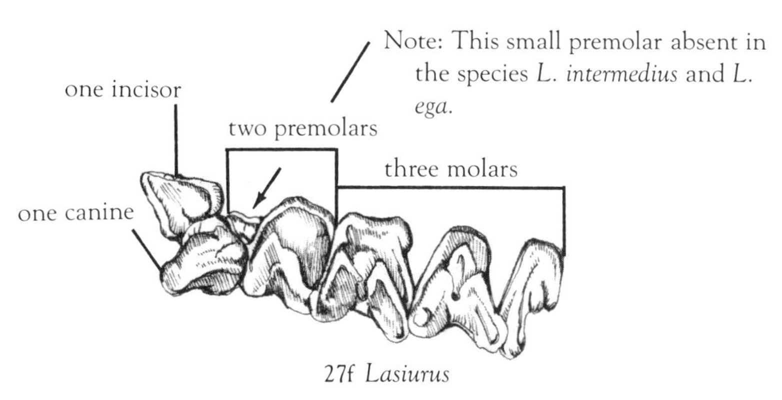

27f *Lasiurus*

Figure 27. (cont.)

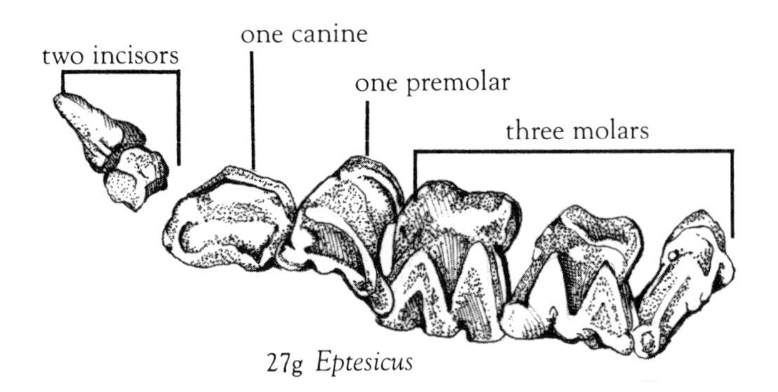

27g *Eptesicus*

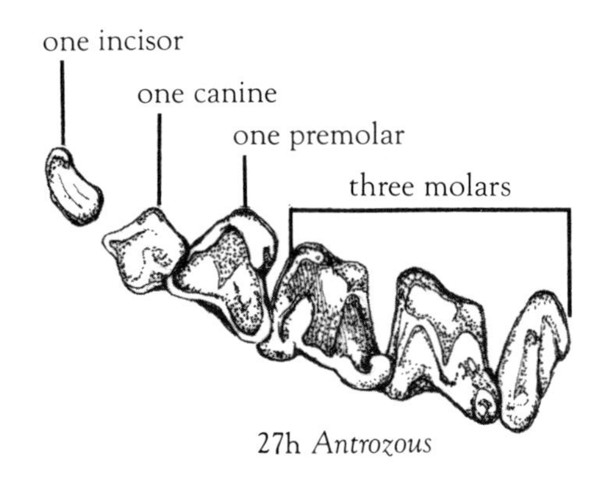

27h *Antrozous*

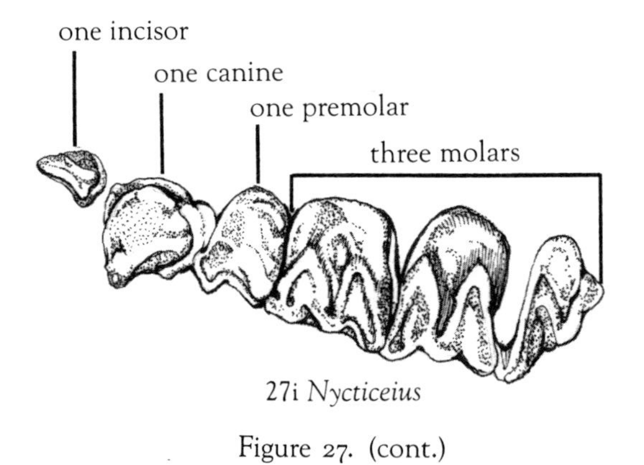

27i *Nycticeius*

Figure 27. (cont.)

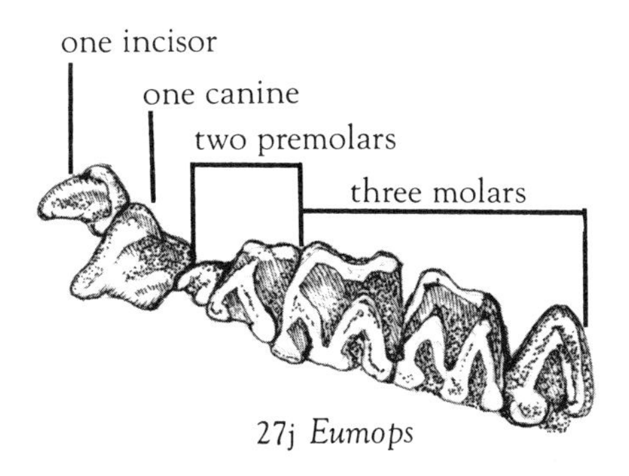

27j *Eumops*

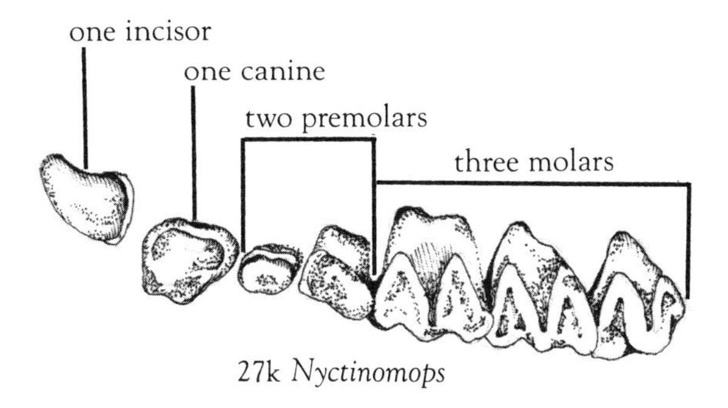

27k *Nyctinomops*

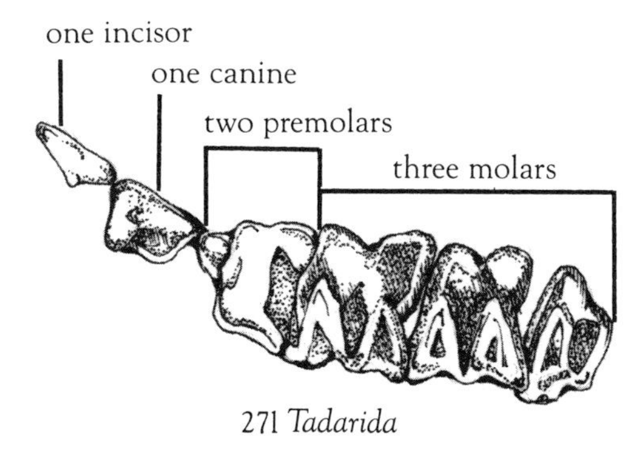

27l *Tadarida*

Figure 27. (cont.)

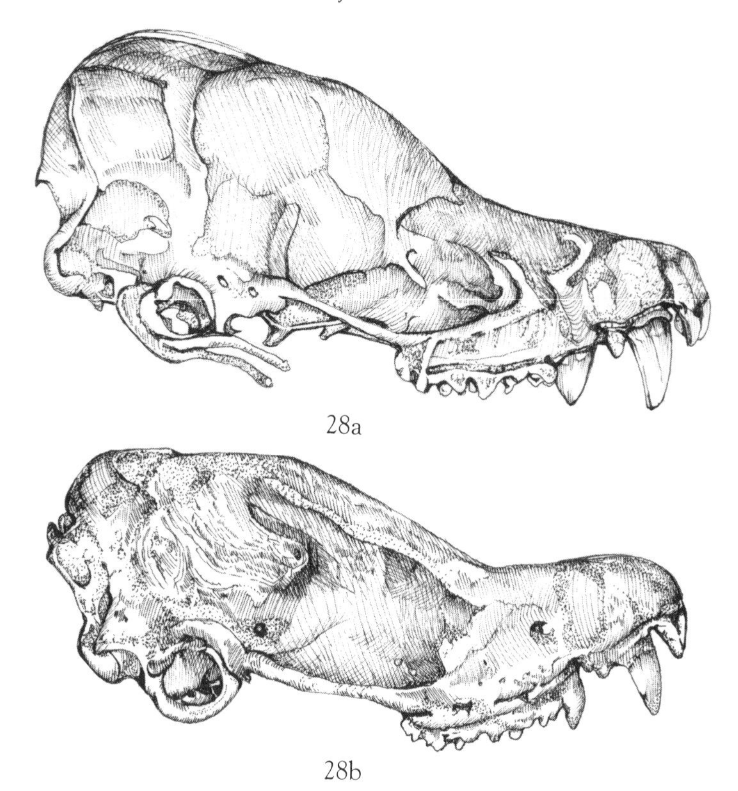

28a

28b

Figure 28.

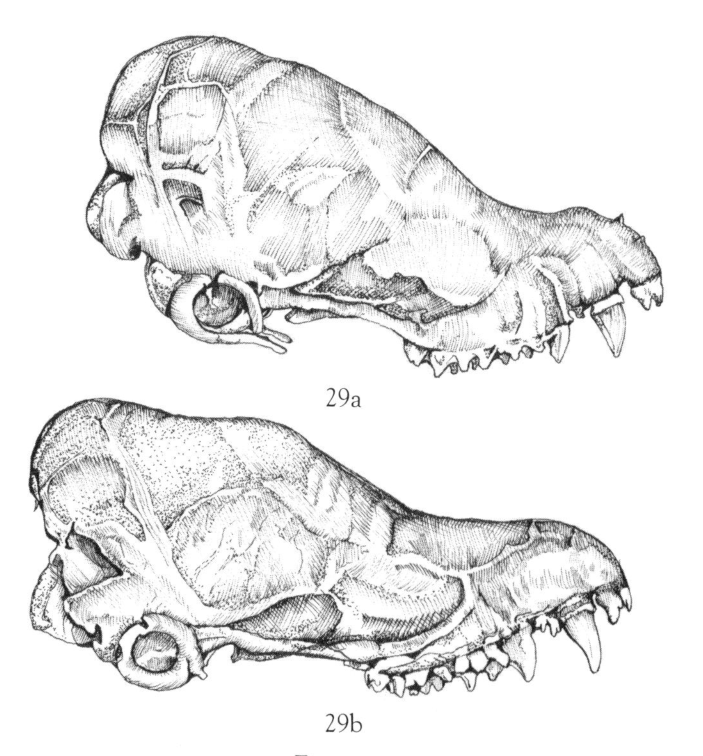

29a

29b

Figure 29.

M. *californicus*

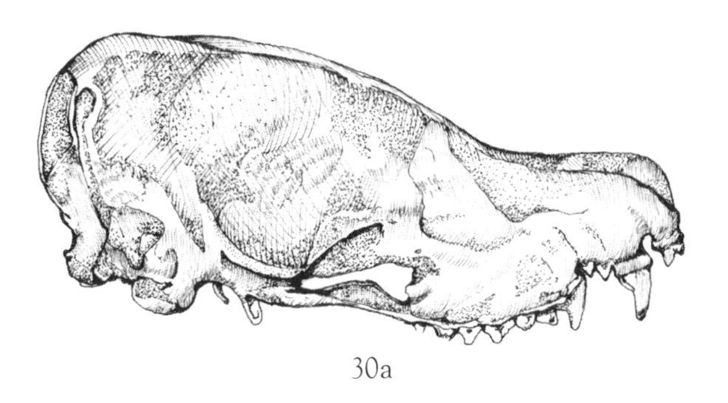

30a

M. *yumanensis*

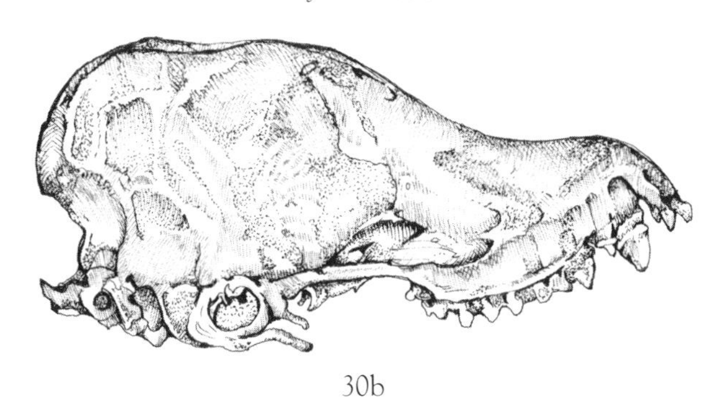

30b

M. *ciliolabrum*

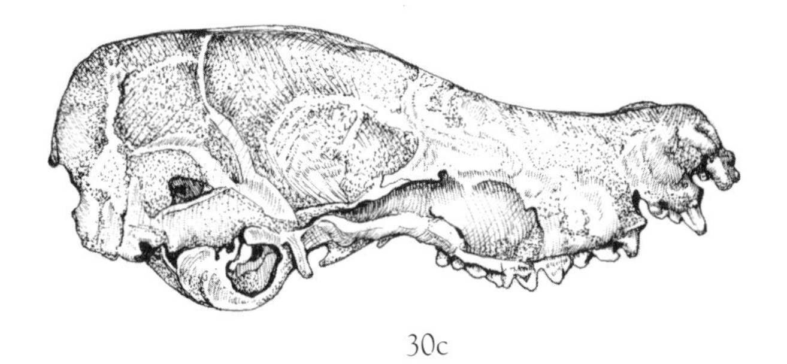

30c

Figure 30.

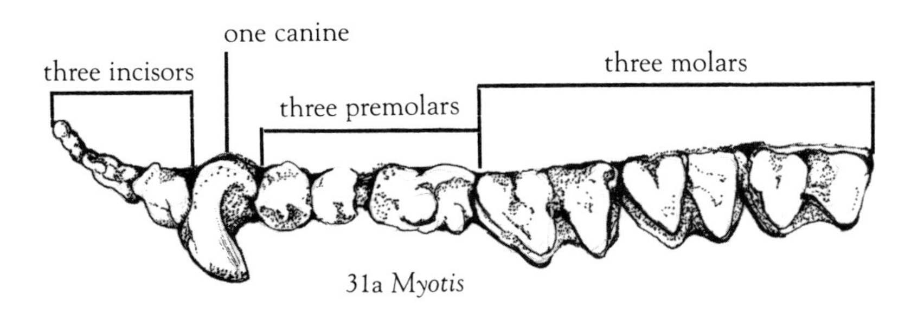

31a *Myotis*

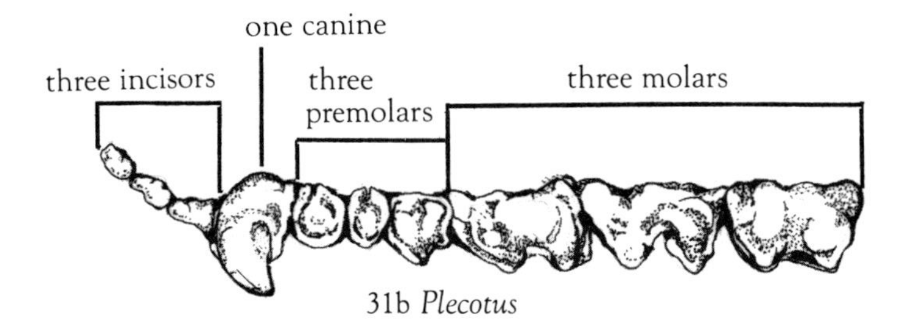

31b *Plecotus*

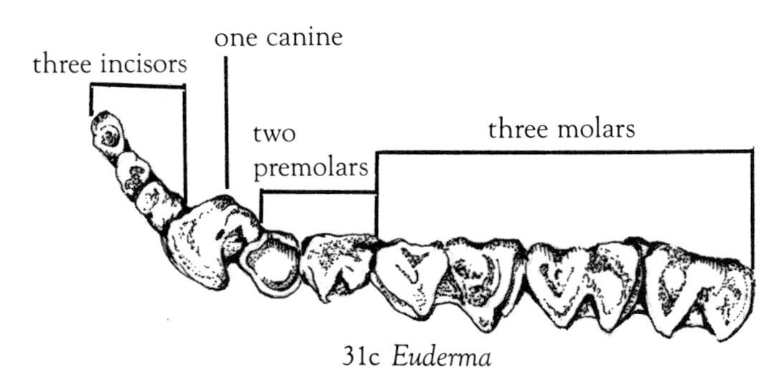

31c *Euderma*

Figure 31. Lower teeth of Texas bats

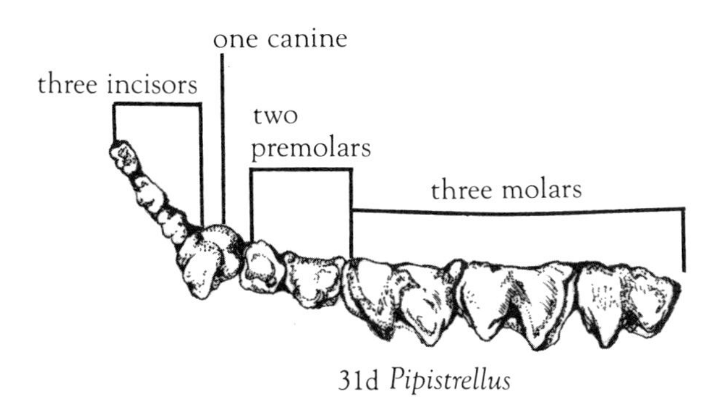

31d *Pipistrellus*

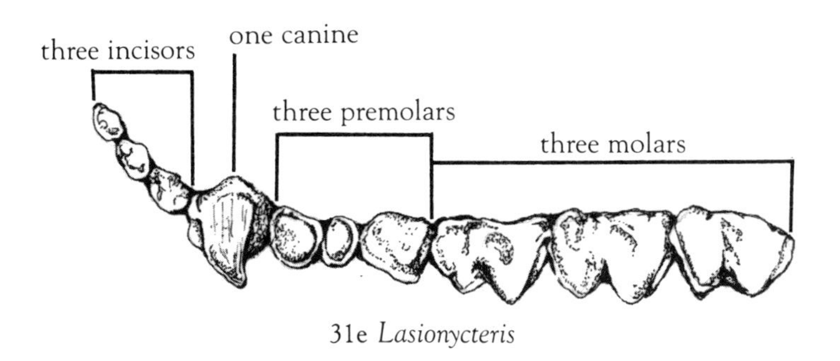

31e *Lasionycteris*

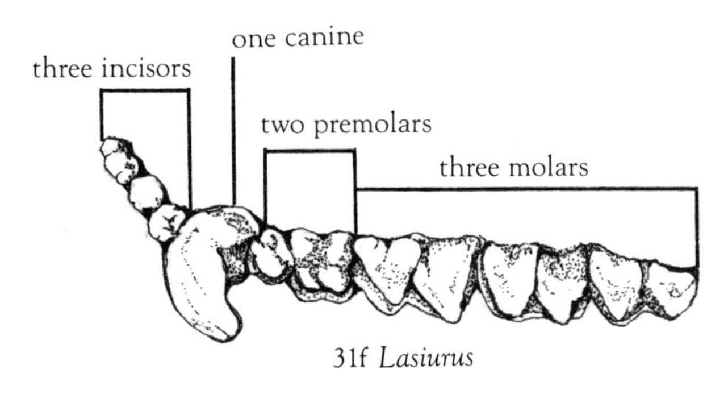

31f *Lasiurus*

Figure 31. (cont.)

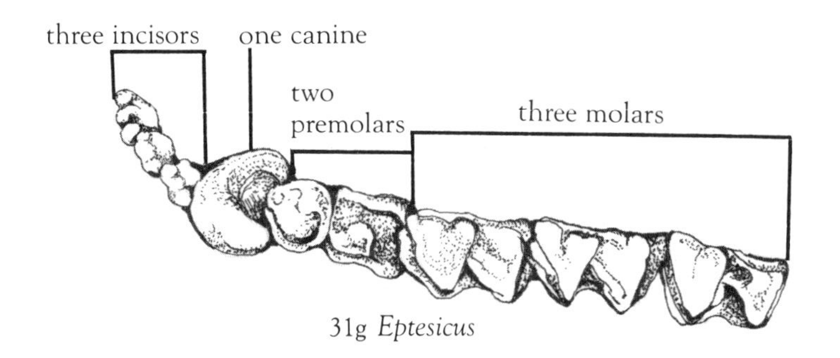

31g *Eptesicus*

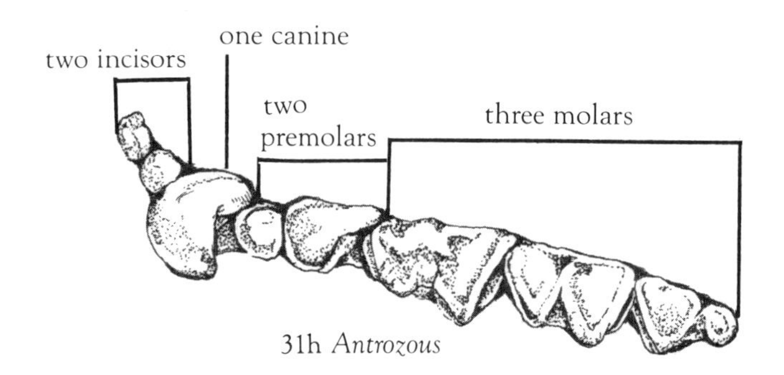

31h *Antrozous*

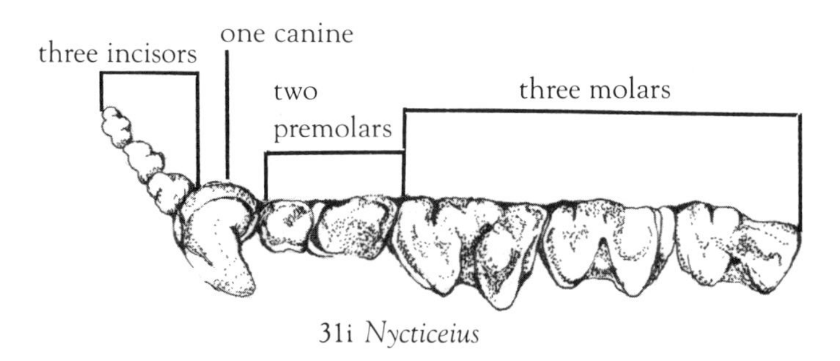

31i *Nycticeius*

Figure 31. (cont.)

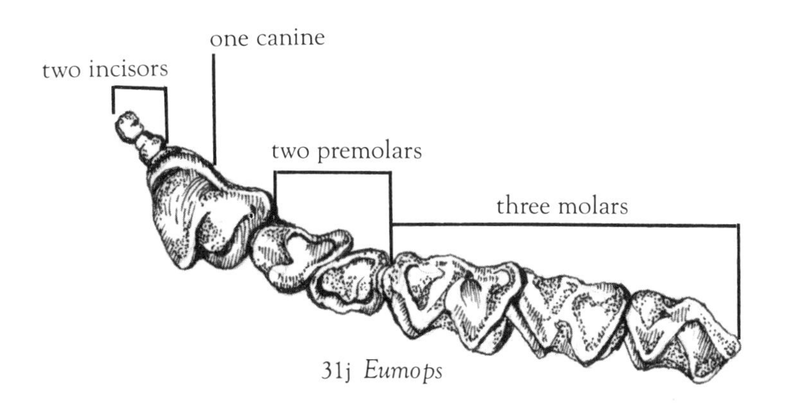

31j *Eumops*

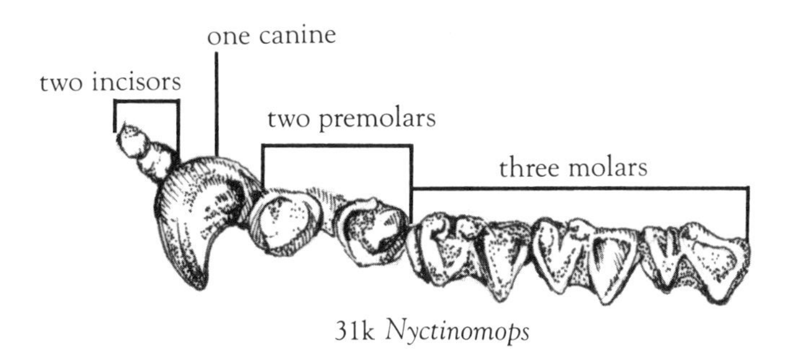

31k *Nyctinomops*

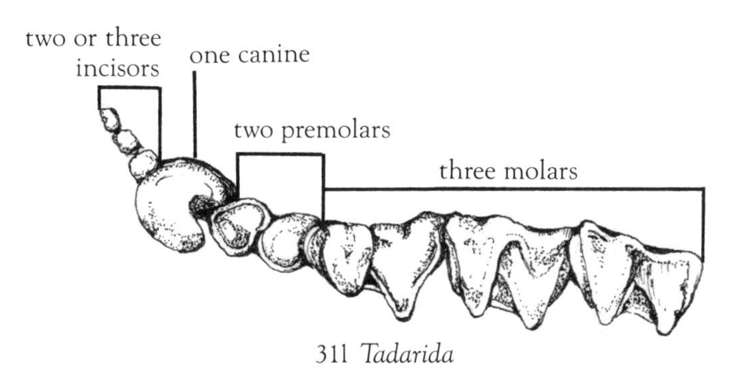

31l *Tadarida*

Figure 31. (cont.)

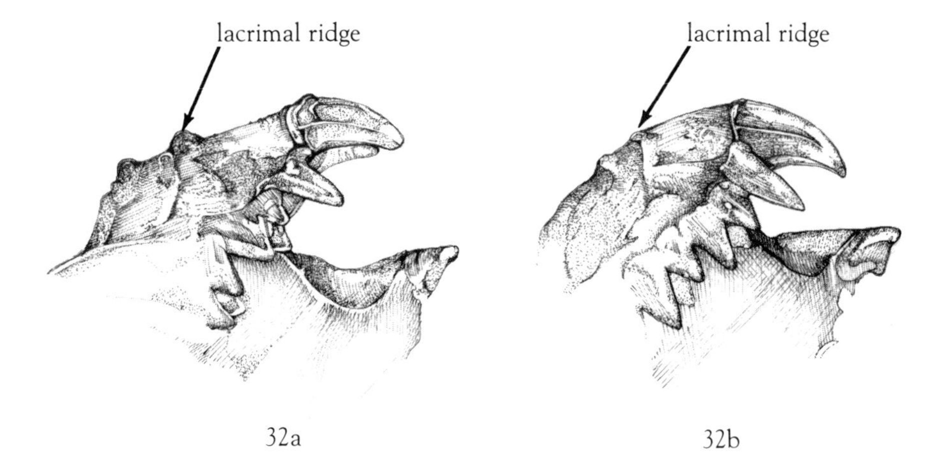

Figure 32.

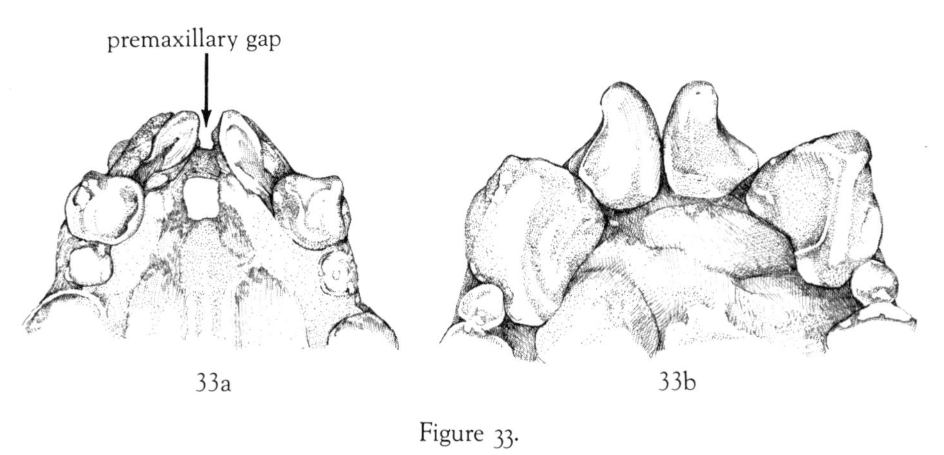

Figure 33.

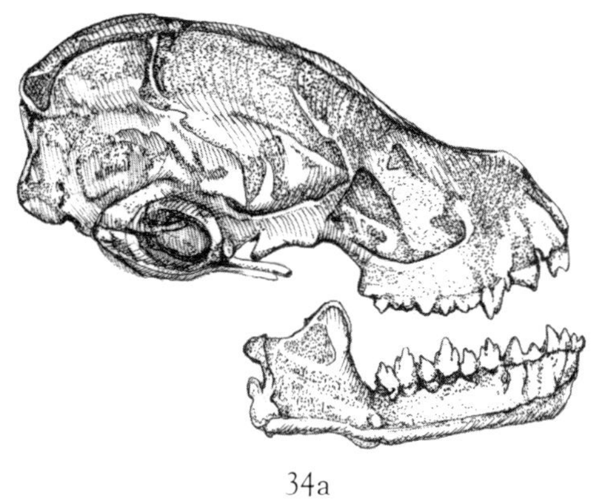

34a

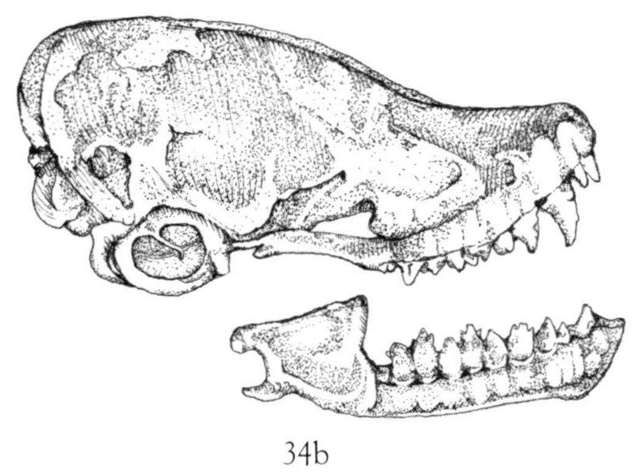

34b

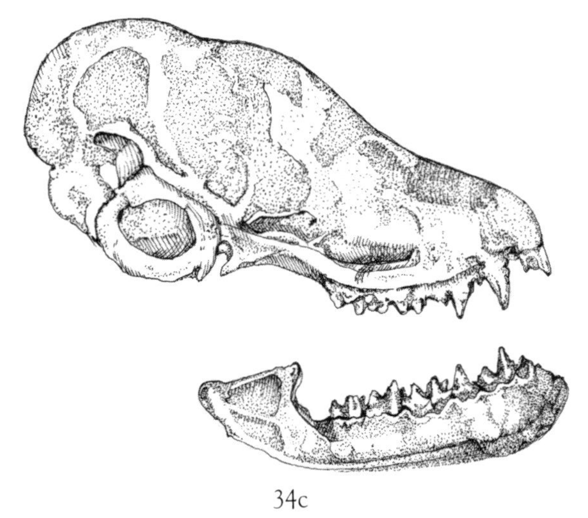

34c

Figure 34.

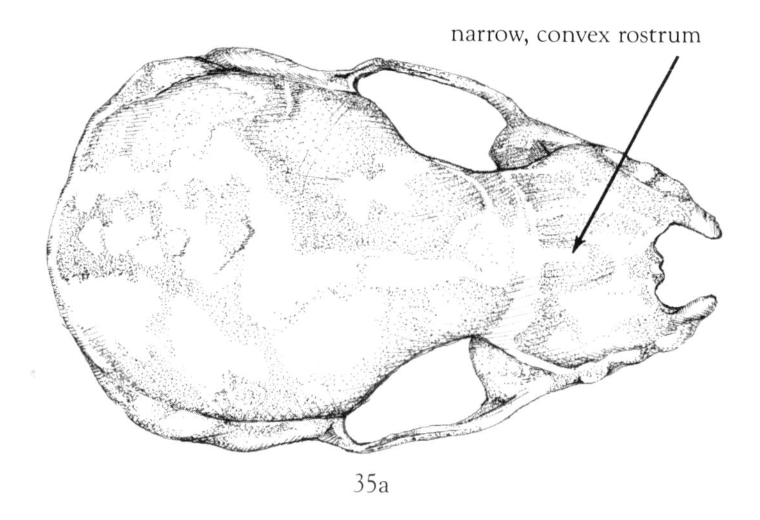

35a

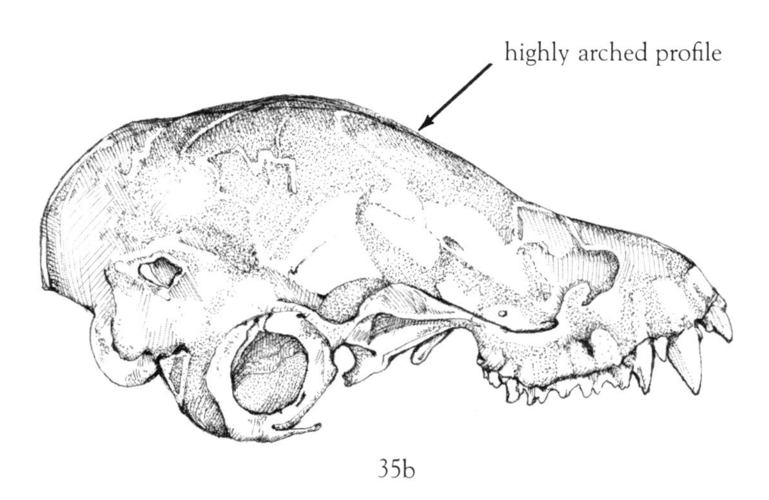

35b

Figure 35.

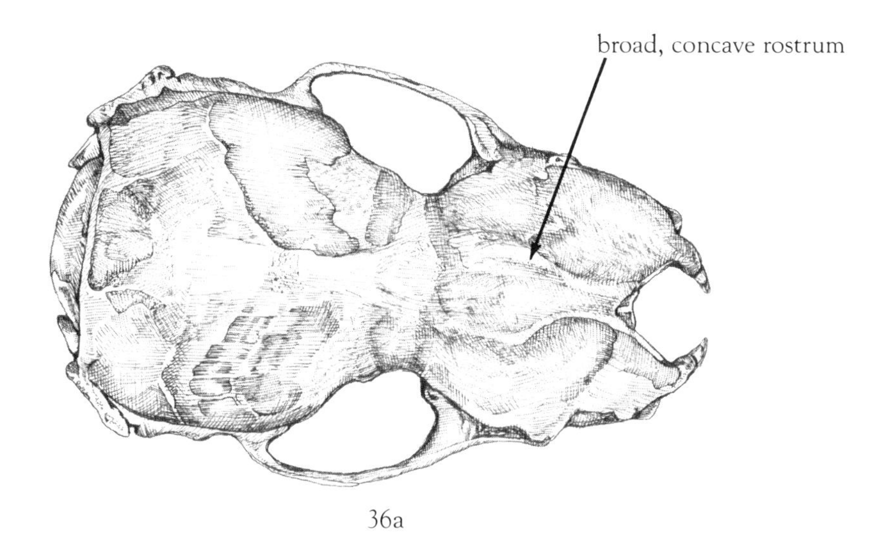

36a

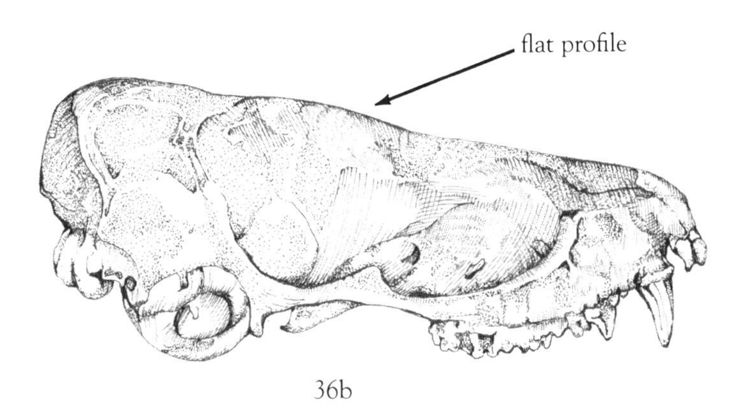

36b

Figure 36.

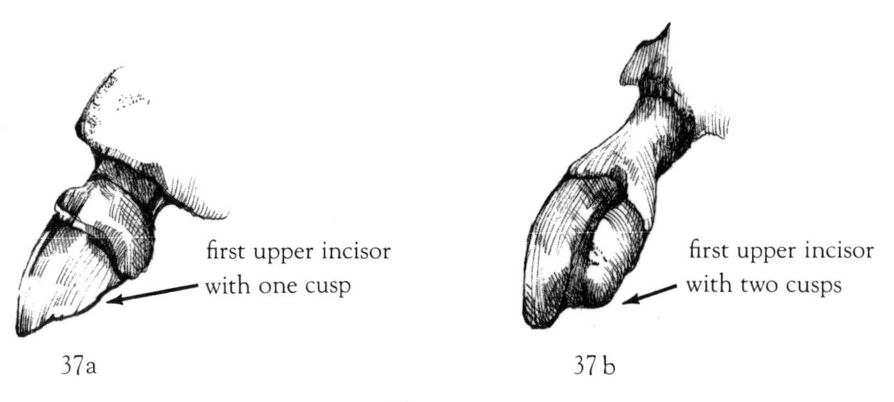
first upper incisor
with one cusp
37a
first upper incisor
with two cusps
37 b

Figure 37.

Choeronycteris mexicana. Mexican Long-tongued Bat. (David J. Schmidly)

Myotis yumanensis. Yuma Myotis. (Merlin D. Tuttle, Bat Conservation International)

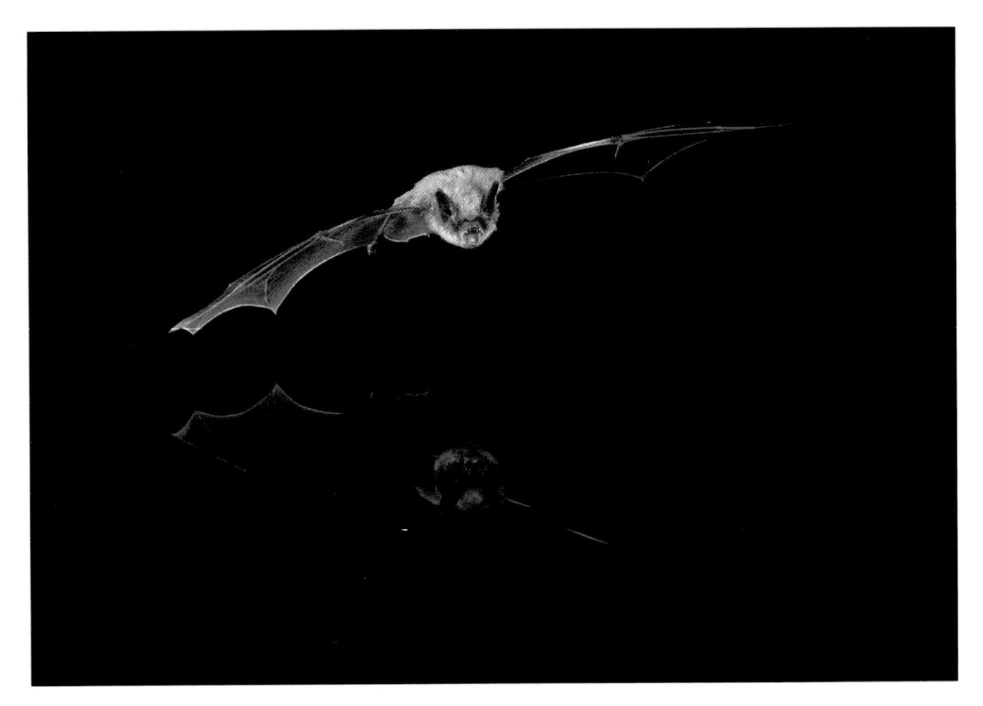

Myotis yumanensis. Yuma Myotis over water. (J. Scott Altenbach)

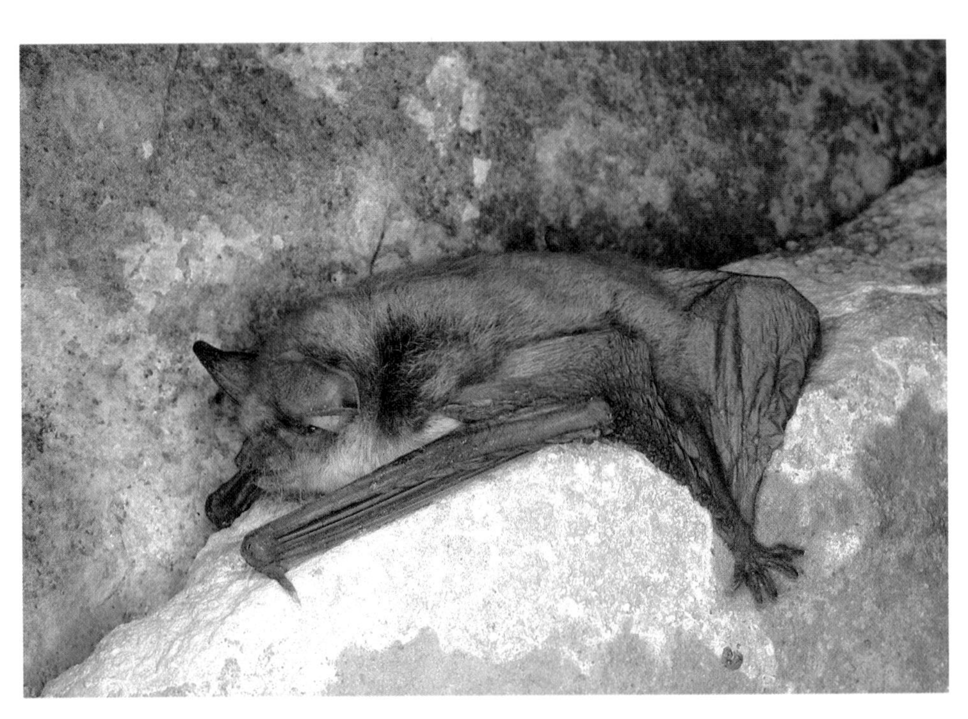

Myotis velifer. Cave Myotis. (John L. Tveten)

Myotis thysanodes. Fringed Myotis. (John L. Tveten)

Eptesicus fuscus. Big Brown Bat. (Merlin D. Tuttle, Bat Conservation International)

Lasiurus borealis. Eastern Red Bat. (Merlin D. Tuttle, Bat Conservation International)

Lasiurus borealis. Eastern Red Bat—mother and young. (John L. Tveten)

Lasiurus cinereus. Hoary Bat. (Merlin D. Tuttle, Bat Conservation International)

Euderma maculatum. Spotted Bat in flight. (Merlin D. Tuttle, Bat Conservation International)

Euderma maculatum. Spotted Bat. (Merlin D. Tuttle, Bat Conservation International)

Plecotus townsendii. Townsend's Big-eared Bat. (Merlin D. Tuttle, Bat Conservation International)

Antrozous pallidus. Pallid Bat in flight. (Merlin D. Tuttle, Bat Conservation International)

Tadarida brasiliensis. Brazilian Free-tailed Bat. (John L. Tveten)

Nyctinomops macrotis. Big Free-tailed Bat. (Merlin D. Tuttle, Bat Conservation International)

Eumops perotis. Western Mastiff Bat. (Merlin D. Tuttle, Bat Conservation International)

3

Accounts of Species

FAMILY MORMOOPIDAE

Eight species of New World bats make up the family Mormoopidae, which includes the mustached, naked-backed, and ghost-faced bats that are variously distributed from the southern United States through Mexico, Central America, and South America to southern Brazil. Characterized by fleshy appendages on the snout and chin and a short tail protruding dorsally from the interfemoral membrane, these bats are abundant in the tropics as well as semi-arid and subtropical environments. The only species of this family to occur in the United States is the ghost-faced bat, *Mormoops megalophylla,* which has been recorded in Texas and Arizona.

Mormoops megalophylla (Peters, 1864)
Ghost-faced Bat

Description. This is a relatively large (forearm=51–59 mm), "pug-faced" bat with reddish brown to dark brown pelage. The ears are large, rounded, and join across the forehead. *Mormoops* is readily recognizable by its distinctive facial ornamentations, sharply upturned rostrum, and tail. Numerous tubercles and folds cover the rostrum and lips, and the ears are shaped and connected in such a way that the animal's mouth, when opened, functions as an orifice at the bottom of this funnel-shaped group of structures. The tail protrudes dorsally from the interfemoral membrane with approximately 1.3 cm not attached to this membrane and lying above it. Average external measurements are: total length, 93 mm; tail, 27 mm; hind foot, 12 mm; ear, 13 mm.

Distribution. In Texas this bat is known from the Trans-Pecos, southern edge of the Edwards Plateau, and South Texas Plains. It is typically found in lowland areas, especially desert scrub and riverine habitats, although it also has been captured in the mountainous country of the Apache, Chisos, and Chinati mountains and in the Sierra Vieja range. It is a common winter (November 1 to March 15) resident of caves along the extreme southern edge of the Edwards Plateau, where

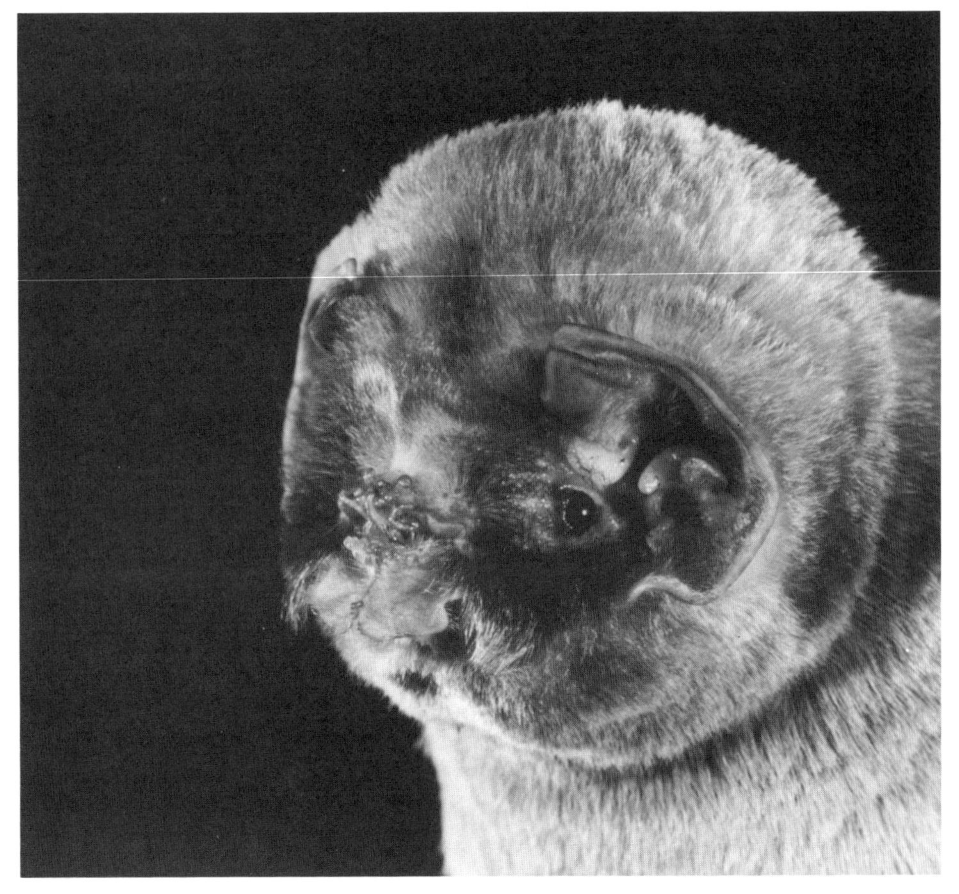

Mormoops megalophylla. Ghost-faced Bat. (Merlin D. Tuttle, Bat Conservation International)

it apparently reaches its northern distributional limits in the United States. However, its occurrence at specific localities is highly variable and unpredictable.

Mormoops has been collected at Frio Cave (Uvalde County), Webb Cave (Kinney County), Haby Cave (Bexar County), and Valdina Farms Sinkhole (Medina County) in December, January, February, March, May, September, and November, indicating that it uses the Edwards Plateau caves as a winter retreat. Many apparently suitable caves near those listed above are not used, although they are occupied by bats often associated with *Mormoops.*

Eads et al. (1956) described the chamber in Frio Cave used by *Mormoops* as "10 to 20 feet [3.0–3.6 m] high, 30 to 40 feet [9.1–12.2 m] wide and about the same length." Temperature and humidity in this chamber, which was called the "warm room," were higher than in the remainder of the cave. As many as six thousand ghost-faced bats were observed at this site during winter, whereas the colony at Haby Cave was described as "small."

In contrast to the winter records from the Edwards Plateau, those from the Trans-Pecos are from the warmer months of the year (March 16 to October 31).

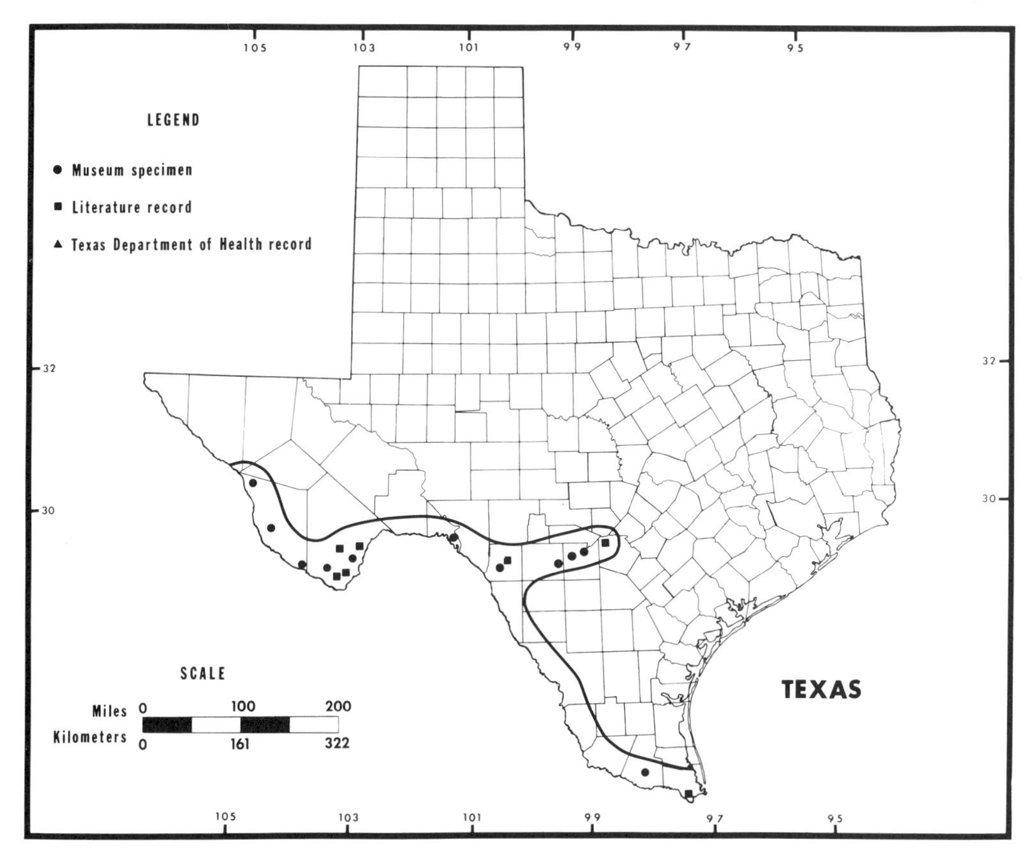

Map 1. Distribution of the ghost-faced bat, *Mormoops megalophylla megalophylla*.

This is suggestive of seasonal migration between the two regions, although such movements have yet to be substantiated.

Subspecies. Texas specimens are referrable to the subspecies *M. m. megalophylla* (Peters, 1864), as indicated by the most recent taxonomic revision of the species (Smith, 1972).

Life History. Ghost-faced bats typically roost in caves, tunnels, and mine shafts, but they also have been found in old buildings. Although they may congregate in large numbers at a roosting site, these bats do not form the compact clusters typical of many other species; rather, they tend to space themselves approximately 15 cm apart across the roost ceiling. Therefore, larger sites are required to house great numbers of these bats. Other species which often cohabit roosting sites with *Mormoops*, and which generally greatly outnumber them, include the cave myotis (*Myotis velifer*) and the Brazilian free-tailed bat (*Tadarida brasiliensis*).

Very little is known about the food habits of ghost-faced bats. Easterla (1973) examined the stomach contents of two individuals from Big Bend National Park and found they were filled entirely with large moths.

Data on the reproductive habits are equally sketchy. One young, born in late

May to early June, is believed to be the norm. Easterla (1973) examined two pregnant females, each containing a single embryo, on June 15, 1968, in Big Bend National Park. Lactating females were recorded from June 15 to August 9.

Specimens examined (46). *Presidio Co.*: ZH Canyon, Sierra Vieja, 3 (TTU); 14 mi E Ruidosa, Pinto Canyon, Chinati Mts., 2 (1 TTU, 1 UIMNH); 30 mi SSE Redford, 4 (MWSU). *Brewster Co.*: Dead Horse Mts., BBNP, 1 (BBNHA); Oak Creek, BBNP, 5,000 ft., 3 (1 BBNHA, 2 TCWC). *Val Verde Co.*: 9 mi W Comstock, 1 (TNHC); 9 mi SW Comstock, 1 (TNHC). *Kinney Co.*: Fort Clark, 1 (USNM/FWS). *Uvalde Co.*: Concan, 5 (1 AMNH, 4 TCWC); Frio Cave, 9 mi S Concan, 17 (13 LACM, 4 TTU); Sabinal, 2 (USNM/FWS). *Medina Co.*: Nye Cave, 1 (LACM); Valdina Farms Sinkhole, 3 (LACM); no specific locality, 1 (KU). *Hidalgo Co.*: Edinburg, 1 (TCWC).

Additional records: *Culberson Co.*: 6 mi N Kent (Dalquest and Stangl, 1986). *Brewster Co.*: BGWMA (Walton and Kimbrough, 1970); W base Chilicotal Mts., BBNP (Constantine, 1961b; Whitaker and Easterla, 1975); Giant Dagger Yucca Flats, BBNP, 3,000 ft. (Easterla, 1968); Boot Springs, BBNP (Easterla, 1973). *Kinney Co.*: Fort Clark (Rehn, 1902); Webb Cave, Shahan Ranch, 7 mi N Bracketville (Raun and Baker, 1959). *Bexar Co.*: Haby Cave (Eads et al., 1956). *Medina Co.*: Valdina Farms Sinkhole, 16 mi N D'Hanis (Raun and Baker, 1959). *Cameron Co.*: Brownsville (Taylor and Davis, 1947).

References. 1, 2, 3, 4, 7, 12, 15, 21, 24, 29, 33, 76, 96, 97, 131, 133, 135, 155, 161, 170, 181, 185, 255, 265, 297, 306, 310, 314, 340, 342, 351, 362, 377, 412, 414, 415.

FAMILY PHYLLOSTOMATIDAE

The Phyllostomatidae are a large, New World family of bats primarily limited to tropical and subtropical areas, although a few species reach northward to subtemperate areas in the United States. The 137 species classified into this family are characterized by a fleshy appendage or "nose leaf" projecting from the rostrum. Most of these bats feed upon fruit or nectar, but this family also contains a few insectivores, carnivores, and the true vampire bats. Three species of phyllostomatid bats have been recorded in Texas, including one vampire, but none is widely distributed or very common.

Choeronycteris mexicana Tschudi, 1844
Mexican Long-tongued Bat

Description. This is a relatively large bat (forearm = 43–45 mm) with a long, slender muzzle and a prominent nose leaf. A minute tail is present and extends less than halfway to the edge of the interfemoral membrane. Color is sooty gray to brownish. Average external measurements are: total length, 85 mm; tail, 10 mm; hind foot, 14 mm; ear, 16 mm.

Choeronycteris mexicana. Mexican Long-tongued Bat. (Merlin D. Tuttle, Bat Conservation International)

This bat is similar to the Mexican long-nosed bat (*Leptonycteris nivalis*). The tail of *L. nivalis,* however, is not evident and the interfemoral membrane is reduced to a mere fringe lining the insides of the legs, in contrast to the better developed interfemoral membrane of *C. mexicana.*

Distribution. The first documented record of the Mexican long-tongued bat in Texas was of a single specimen captured on December 11, 1970, at Santa Ana National Wildlife Refuge, Hidalgo County, in extreme South Texas. This bat was not preserved and is known only from photographs deposited in the Texas Cooperative Wildlife Collection at Texas A&M University; however, the photographs clearly indicate the bat was *C. mexicana.* Subsequently, other specimens have been obtained at Santa Ana Refuge but they also were not preserved (Fred Gehlbach, pers. comm.).

Subspecies. *C. mexicana* Tschudi, 1844, is a monotypic species with no subspecies recognized.

Life History. Very little is known about the biology of this bat, which also occurs as a summer resident of southeastern Arizona and extreme southwestern New Mexico. The record from Texas is of interest because it suggests that the species may overwinter in our state. Mexican long-tongued bats are known to inhabit deep canyons in the small, insular mountain ranges where they use caves and mine tunnels as day roosts. These bats have also been found in buildings and often occur in association with Townsend's big-eared bats (*Plecotus townsendii*).

Mexican long-tongued bats feed primarily on fruit, pollen, nectar, and probably insects. Because of their longer tongues, they may be able to recover nectar from a greater variety of night-blooming plants than the other nectar-feeding bat occurring in Texas, *Leptonycteris nivalis*.

Parturition occurs from June to early July in Arizona and New Mexico with young reported as early as mid-April in Sonora, Mexico. The litter size is one. I collected in May a pregnant female (which gave birth shortly after capture) in the San Carlos Mountains of northern Tamaulipas, Mexico, which is no more than 240 km south of Santa Ana Refuge. Pregnant and lactating females have been recorded in March and June in Coahuila, Mexico, south of the Texas border (Wilson, 1979).

Specimens examined (0).

Additional record: *Hidalgo Co.*: Santa Ana National Wildlife Refuge (LaVal and Shifflet, 1971).

References. 2, 3, 4, 10, 12, 15, 21, 33, 37, 79, 135, 190, 245, 265, 281, 329, 362.

Leptonycteris nivalis (Saussure, 1860)
Mexican Long-nosed Bat

Description. This is a relatively large (forearm = 46–59 mm), sooty brown bat with a long muzzle and prominent nose leaf. The tail is minute and appears to be lacking but actually consists of three vertebrae, and the interfemoral membrane exists as a narrow fringe along the inside of each leg. *Choeronycteris mexicana* is similar in appearance, but has a broader interfemoral membrane, with a distinct tail and a longer, narrower muzzle. Average external measurements of *L. nivalis* are: total length, 83 mm; hind foot, 15 mm; ear, 15 mm.

Distribution. A highly colonial, cave-dwelling, migratory species, this bat has been recorded in the United States only in Trans-Pecos Texas, where it has been captured in Big Bend National Park, Brewster County, and the Chinati Mountains of Presidio County. Although occasionally found in the desert scrub habitats at lower elevations, these bats apparently prefer mountainous, pine-oak habitats at elevations of 1,500 to 2,300 m. Mexican long-nosed bats occupy such areas from June to August, after which they move out of the state to winter in Mexico. Few adult males have been recorded in Texas, which suggests that the sexes may segregate geographically with males rarely appearing in the most northerly part of the species' range.

Although this bat occurs throughout much of Mexico, it is relatively rare over most of its range and there are indications of substantial population decline. Several caves in central Mexico known to house considerable numbers of these bats in the past now contain only small colonies or lack bats altogether. A colony of Mexican long-nosed bats found in the Chisos Mountains of Big Bend National Park (Emory Peak Cave) is the only known population of this species in

Leptonycteris nivalis. Mexican Long-nosed Bat. (Merlin D. Tuttle, Bat Conservation International)

the United States. First discovered in 1937 (Borell and Bryant, 1942), this colony fluctuates widely in numbers from one year to the next, with yearly estimates of population size ranging from zero to as many as 13,650 individuals. Reasons for this instability are unknown, but it has been suggested that the colony forms in years when overpopulation or low food supply in Mexico forces the bats to move northward (Easterla, 1972a). Considering this information, the U.S. Fish and Wildlife Service added this bat, along with its close relative *L. sanborni*, to the list of endangered species in 1988. Both species are jeopardized by disturbance of roosting sites, loss of food sources, and direct killing by humans.

Subspecies. *L. nivalis* (Saussure, 1860) is a monotypic species and no subspecies are recognized.

Life History. These bats typically cluster to roost in caves where they may occasionally be found in association with Townsend's big-eared bats (*Plecotus townsendii*). They are exceptionally strong, highly maneuverable fliers, and are able to hover in flight while they feed, much as hummingbirds do.

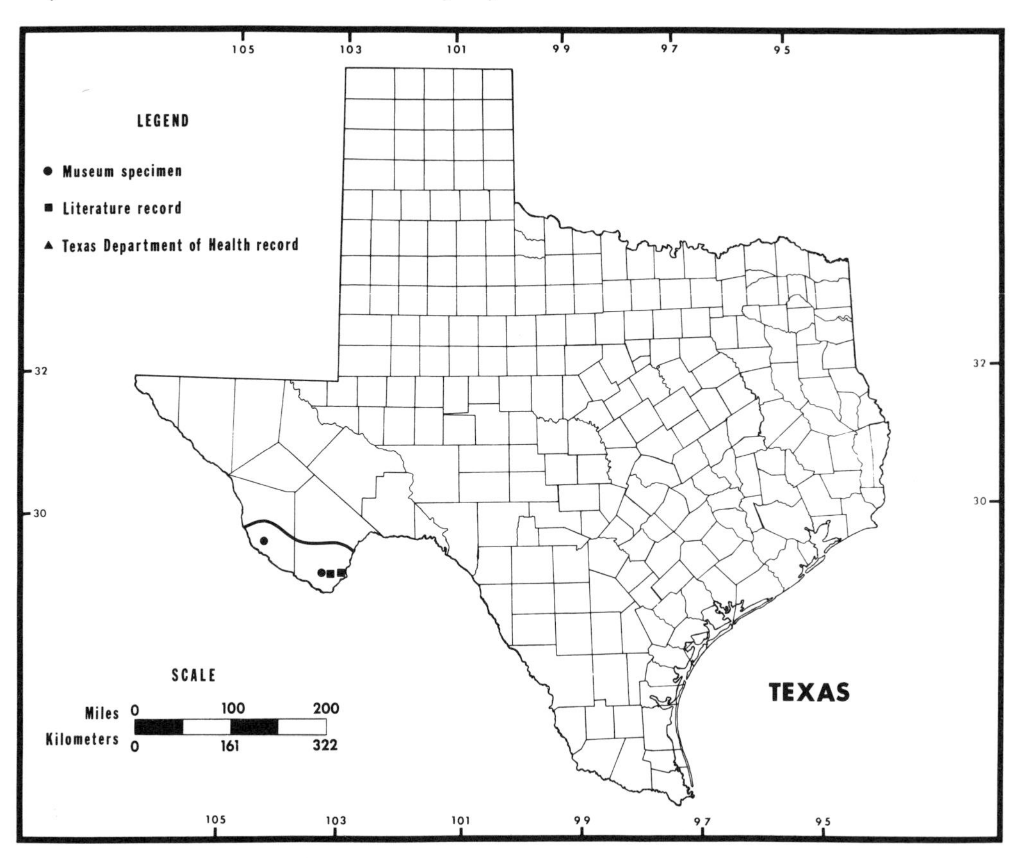

Map 2. Distribution of the Mexican long-nosed bat, *Leptonycteris nivalis.*

Nectar and pollen are the main food items for long-nosed bats, with the flowers of century plants (*Agave* spp.) probably their main food source in Texas. Their seasonal occurrence in Texas is probably related to food availability, as their presence seems to coincide with the blooming of century plants in June. These plants open their flowers at night and attract bats with copious amounts of nectar. As the bats feed, their fur gets coated with pollen grains. When they fly to another plant in search of more food, they transfer the pollen to a new flower, assisting in cross-fertilization of the plants. Both the plant and the bat benefit from this mutual relationship. This association has resulted from the coevolution of bats and plants, and the dependence is so strong that the plants could not reproduce without the intervention of the bats. Likewise as the flower stalks of the agaves die by late summer, the bats disappear from the region because there is nothing left for them to eat.

Phillips et al. (1969) reported the occurrence in the oral cavity of *L. nivalis* of mite (genus *Radfordiella*) infestations, which apparently are responsible for dental lesions along the first and second premolars, and occasionally the molars of the upper toothrow. In 13.6 percent of specimens of *L. nivalis* having lesions, one or more teeth had been lost as a result of destruction of dental and supporting tissues. Oral mites have not been found in *L. sanborni*, a bat closely related to, and in part sympatric with *L. nivalis*, and the presence or absence of oral mites is a useful taxonomic character in distinguishing between these two species (Greenbaum and Phillips, 1974; Phillips et al., 1969).

Available data indicate that the breeding season is restricted to April, May, and June (Wilson, 1979). The young probably are born in Mexico, prior to the time bats arrive in the Big Bend area of Texas. Females are believed to give birth to one or perhaps two young annually. The young are weaned in July or August, which is the peak of the rainy season and the period of peak flower abundance. In September or October, individuals migrate back to Mexico for the winter.

Remarks. Another species of *Leptonycteris*, *L. sanborni*, has been recorded from southwestern New Mexico and northern Mexico, and this species could occur in Texas, although it has not been recorded in our state to date. The two species can easily be distinguished by characters of the pelage. *L. nivalis*, compared with *L. sanborni*, is slightly larger (forearm, 55–60 mm vs. 52–56 mm); has a longer third finger (106–15 mm vs. 92–102 mm); has a conspicuous fringe of hair on the uropatagium; and has long, fluffy hair in contrast to the short, dense hair of *L. sanborni*.

Specimens examined (94). *Presidio Co.*: 14 mi E Ruidosa, Pinto Canyon, Chinati Mts., 4 (TTU). *Brewster Co.*: 0.25 mi above Boot Springs, BBNP, 2 (BBNHA); Basin Sewage Lagoon, BBNP, 1 (MSB); Emory Peak Cave, BBNP, 2 (1 BBNHA, 1 MSB); Emory Peak, BBNP, 7,000–7,500 ft., 85 (57 FMNH, 6 MVZ, 22 TCWC).

Additional records: *Brewster Co.*: Ernst Canyon, BBNP (Easterla, 1973); Boot Springs Lake, Pine Canyon Tank, BBNP (Easterla, 1973); Basin Sewage Lagoon, BBNP (Easterla, 1973); Emory Peak Cave, BBNP (Easterla, 1973); Rio Grande Village *Gambusia* pools, BBNP (Easterla, 1973).

References. 1, 2, 3, 4, 7, 12, 21, 24, 29, 33, 45, 79, 81, 105, 135, 162, 179, 181, 211, 244, 265, 306, 310, 329, 330, 342, 362, 373, 384, 414, 421, 422.

Diphylla ecaudata Spix, 1823
Hairy-legged Vampire

Description. This is a relatively large (forearm = 53 mm), dark brown bat with short, rounded ears and a short, "pug-nosed" snout. As with other phyllostomatid bats, the hairy-legged vampire also has a nose leaf, although this feature is much reduced. The tail is absent and a narrow, hairy interfemoral membrane extends from the body down the leg. The dentition is highly modified with the upper incisors and canines enlarged and blade-like; premolars and molars are much reduced. Average external measurements are: total length, 83 mm; hind foot, 18 mm; ear, 16 mm.

The hairy-legged vampire is most easily confused with the common vampire, *Desmodus rotundus*, from which it may be distinguished by characters of the skull and pelage. In *Diphylla* the first lower incisors contact and the interfemoral membrane is lined with a distinct fringe of moderately long hairs. In *Desmodus*, the

Diphylla ecaudata. Hairy-legged Vampire. (Merlin D. Tuttle, Bat Conservation International)

first lower incisors are not in contact and the interfemoral membrane lacks a fringe of hair.

Distribution. Primarily tropical in distribution, this bat is known from Texas on the basis of a single female specimen collected on May 24, 1967, in an abandoned railroad tunnel west of Comstock in Val Verde County. This record extended the range of the hairy-legged vampire approximately 725 km to the northwest of Tamaulipas, Mexico, where it is more often encountered. A thorough search of the caves along the Rio Grande in western Texas may produce additional records of this species.

Subspecies. *D. ecaudata* Spix, 1823, is a monotypic species and subspecies are not recognized.

Life History. Only three species of true vampires exist in the world, and all are known from the New World tropics and subtropics. These are the common (*Desmodus rotundus*) and white-winged (*Diaemus youngii*) vampires in addition to the hairy-legged variety. Most research concerning these unusual bats has been conducted with the common vampire, and very little information is available for the other two species.

All three vampire species occur throughout much of Mexico and range into Central and South America, although the white-winged and hairy-legged species are less frequently encountered. Unlike the common vampire, which feeds upon the blood of other mammals—including humans on occasion—white-winged and hairy-legged vampires primarily attack birds.

Diphylla possesses all of the morphological adaptations for a blood diet that are exhibited by the other two vampires. These include enlarged and razor-sharp, blade-like upper incisors and canines designed for inflicting a small, V-shaped wound on the prey animal. The tongue is modified with small grooves for lapping blood from the wound and an anticoagulant agent in the bat's saliva functions to keep the wound open and blood flowing.

Further accommodating their peculiar diet, the digestive tract has been highly modified. The stomach walls are thin and elastic, allowing a large volume of blood to be ingested at one meal (approximately 30 cm^3 in the common vampire), and a network of capillaries around the stomach allows for rapid absorption and excretion of water contained in the blood meal. Although they are exceptionally agile when on the ground, vampires have difficulty flying after feeding and begin to urinate soon after commencing their meal, in an effort to reduce their weight.

Vampires roost primarily in caves, although they have also been found in tunnels and hollow trees. They commonly roost in association with other bats, many of which are economically important insectivores and plant pollinators. Vampire bats are thought to be reproductively active throughout the year. They normally have one young per year, although occasionally they may have two. There is no indication that *Diphylla* breeds in Texas. It appears to have a well-defined breeding season, with a single young per year, in northeastern Mexico, where pregnant and lactating females have been recorded in Tamaulipas in November (Wilson, 1979).

Due to their unusual feeding habits, vampire bats (especially *Desmodus rotundus*) are regarded as a serious problem in much of Latin America, because of the damage done to livestock herds through the transmission of rabies. Additionally, infection is possible at wound sites and the anticoagulant agent used by the bats causes further complications as blood continues to flow after the bat has left. Small animals attacked by several bats could conceivably die from blood loss.

Several methods have been tried in attempts aimed at controlling the bats, including the use of dynamite and poison gas at roosting sites. Their tendency to roost with beneficial species makes such practices unwise, however, and alternate methods of control have been suggested. These include chemical anticoagulant injections into the rumens of bovine livestock. After ingestion, this anticoagulant causes internal hemorrhaging in the bats, thus destroying specific, problem-causing vampires. Keeping domestic fowl in roofed wire pens has been suggested as a way to prevent attacks by white-winged and hairy-legged vampires.

It is ironic to note that the colonization of Latin America by Europeans, and their introduction of livestock, probably contributed to an increase in vampire numbers. The hairy-legged vampire, however, certainly remains one of the rarest bats in Texas.

Remarks. Although the common vampire (*Desmodus rotundus*) has not been recorded from Texas during Recent times, it is known from sub-Recent deposits in the Trans-Pecos and specimens have been collected no more than 193 km south of the Rio Grande in northern Tamaulipas, Mexico. Thus, it is distinctly possible that an occasional common vampire may periodically wander into the southern part of our state.

Specimens examined (1). *Val Verde Co.*: 12 mi W Comstock, 1 (TTU).

References. 1, 7, 12, 21, 24, 33, 44, 79, 143, 150, 265, 343.

FAMILY VESPERTILIONIDAE

Chiefly insectivorous bats, the Vespertilionidae constitute the largest family of bats (approximately 300 species) and are distributed worldwide with the exception of arctic regions. Consequently, these bats are found in nearly every conceivable habitat from tropical forests to desert and temperate regions. Many are highly migratory and traverse great distances between summer and winter ranges. Others, however, do not migrate and, instead, hibernate on summer ranges.

Vespertilionids lack the facial adornments found in other families of bats and are often referred to as "plain-faced" bats. Several species have extremely large and complex ears but most have small, simple ears. These bats typically have small eyes and a long tail completely enclosed by a well developed interfemoral membrane.

Thirty-one species of vespertilionid bats range across the United States; of these, twenty-four are known from Texas.

Myotis lucifugus (Le Conte, 1831)
Little Brown Myotis

Description. This is a medium-sized (forearm=34–41 mm; weight=8 g) *Myotis* with large feet and relatively small ears. The pelage has a slight sheen and is tan to dark brown dorsally while the underparts are somewhat paler. The calcar is not keeled. Average external measurements are: total length, 84 mm; tail, 34 mm; hind foot, 11 mm; ear, 14 mm.

This bat may be confused with the Yuma myotis (*M. yumanensis*), which is a smaller species with pale brown to yellowish pelage and which has smaller and lighter colored ears (table 8). The cave myotis (*M. velifer*) is larger and usually darker than *M. lucifugus*. The California myotis (*M. californicus*) and the western small-footed myotis (*M. ciliolabrum*) are both smaller and have a keeled calcar. The long-legged myotis (*M. volans*) is also similar but has a keeled calcar and the underside of the wing is furred to the elbow and knee.

Distribution. Only a single specimen of *M. lucifugus* has been collected in Texas, making this one of the rarest of Texas bats. This specimen, consisting of a skin and skull only, was collected at a locality near Fort Hancock, Hudspeth County, in the Trans-Pecos. It was probably a migrant individual, and it is doubtful that a resident population of this bat occurs in Texas. *M. lucifugus* is commonly en-

Myotis lucifugus. Little Brown Myotis. (Roger W. Barbour)

countered throughout the rest of the United States except for the Great Plains region.

Subspecies. The specimen taken near Fort Hancock is referrable to the subspecies *M. l. occultus* Hollister, 1909. As discussed in the remarks section of this account, some authorities regard *occultus* as specifically distinct from *lucifugus.*

Life History. The eastern subspecies of the little brown myotis (*M. l. lucifugus*) has been studied in more detail than most other bats due to its wide range in North America. Western subspecies, however, are poorly known and little life history information is available.

Hoffmeister (1986) states that this myotis is usually found in ponderosa pine or pine-oak woodland in Arizona, but Findley et al. (1975) indicate that vegetation zone seems unimportant in determining its distribution in New Mexico. Fort Hancock, Hudspeth County, where the bat has been taken in Texas, is in a region of desert scrub vegetation.

In the northeast, *M. lucifugus* may migrate more than 300 km between winter and summer ranges, but in the west they are believed to hibernate near their summer range. During migration these bats may be found in rock fissures, buildings, under bridges or in just about any other crevice or space which will afford shelter (Fenton and Barclay, 1980).

M. lucifugus prey heavily on aquatic insects and often vary their hunting patterns over the course of an evening (Fenton and Barclay, 1980). Initially, they may

TABLE 8. Trenchant morphological characters useful in distinguishing Texas species of *Myotis*

Species	Ear size (usual length)	Hind foot (usual length)	Overall size (forearm length)	Pelage coloration	Special Hair on body	Keel on calcar	Elevation from rostrum to braincase	Sagittal crest
M. *austroriparius*	intermediate (14–16)	large (9–12)	medium (33–39)	dull gray or orange-brown	–	none	abrupt	slight
M. *californicus*	small (9–13)	small (6–8)	small (29–34)	light brown	–	well developed	abrupt	none
M. *ciliolabrum*	small (12–14)	small (6–8)	small (31–36)	buff brown	–	well developed	gradual	none
M. *lucifugus*	small (11–16)	large (9–11)	medium (34–41)	shiny, dark brown	–	none	gradual	distinct
M. *septentrionalis*	large (17–19)	large (9–11)	medium (35–39)	dull, gray-brown	few, isolated hairs on edge of tail membrane	none	gradual	none
M. *thysanodes*	large (16–20)	large (9–12)	large (40–46)	buff brown	thick fringe of hair on edge of tail membrane	none	abrupt	slight
M. *velifer*	intermediate (15–17)	large (9–12)	large (40–44)	dark, dull	bare patch on back between scapulae	none	abrupt	well developed
M. *volans*	small (12–14)	large (9–11)	medium (38–42)	dark brown	underside of wing furred to elbow	well developed	abrupt	none
M. *yumanensis*	small (12–14)	large (9–11)	small (32–36)	light, buff brown	–	none	abrupt	none

feed along the margins of streams and lakes approximately 2 to 5 m above ground; later they forage in groups directly over the water surface, often within 1 m of the surface.

Hoffmeister (1986) indicates that young are born in late June in Arizona. On June 29, 1960, Hayward (1963) discovered a maternity colony of these bats in an abandoned house near the Verde River in Arizona. The river is located in a half-mile-wide valley and is lined with cottonwoods, sycamores, and willows. Sixty-seven adult females were captured and banded but no adult males were found. Most females had already given birth. This site was revisited on June 4, 1961 and although 41 adult females were seen, there were no young present.

Hayward (1963) found the weight of the young to range from 1.8 g (individual almost hairless) to 6.6 g (individual not able to fly as yet). Juvenile males averaged almost 0.5 g heavier than the juvenile females. Other species of bats present were the cave myotis (*Myotis velifer*) and Yuma myotis (*Myotis yumanensis*).

Remarks. Considerable confusion exists concerning the taxonomy of this bat. *Myotis lucifugus* and *M. occultus* were regarded as distinct species until the work of Findley and Jones (1967) and Barbour and Davis (1970) showed that considerable intergradation in cranial and dental characters exists in bats from the southwestern United States. Subsequently, *M. occultus* was grouped with *M. lucifugus* as the subspecies *M. l. occultus* (Hall, 1981).

Recent authors (Findley, 1972; Hoffmeister, 1986) remain unsure of this arrangement, however, and continue to regard the two taxa as separate species. Hoffmeister (1986) re-examined specimens used to unite the two species (Barbour and Davis, 1970) but was unable to conclude that the two are conspecific. He tentatively gave specific status to *M. occultus*.

Specimens examined (1). *Hudspeth Co.*: Fort Hancock, 1 (USNM/FWS).

References. 2, 4, 7, 10, 12, 13, 21, 22, 24, 26, 27, 29, 31, 39, 78, 84, 86, 92, 102, 123, 142, 154, 188, 191, 192, 194, 198, 214, 220, 222, 243, 262, 265, 304, 306, 309, 310, 316, 358, 363, 423.

Myotis yumanensis (H. Allen, 1864)
Yuma Myotis

Description. This is a small (forearm = 32–38 mm), buff brown bat with relatively large feet and short ears. The underparts range from white to buff brown and the calcar is not keeled. Other *Myotis* species sympatric with this one include *M. velifer*, which is larger and darker in coloration; *M. volans*, *M. californicus*, and *M. ciliolabrum*, all of which have keeled calcars; and *M. lucifugus*, which has dark brown, glossy fur (table 8). Average external measurements of *M. yumanensis* are: total length, 80 mm; tail, 34 mm; hind foot, 8 mm; ear, 14 mm.

Distribution. The Yuma myotis is a summer resident of the southern Trans-Pecos region and just east of the Pecos River in Val Verde County; in the Texas Natural

Myotis yumanensis. Yuma Myotis. (Merlin D. Tuttle, Bat Conservation International)

History Collection there is also an unreported specimen from Starr County in the South Texas Plains region. This bat is most commonly encountered in lowland habitats near open water where it prefers to forage. Most specimens collected in Texas have come from areas near the Rio Grande. Although no records are available, this species likely hibernates in the Trans-Pecos during winter months.

Subspecies. Texas specimens are referrable to the subspecies *M. y. yumanensis* (H. Allen, 1864), based on the most recent taxonomic revision of the species (Harris, 1974).

Life History. As with the little brown bat, the Yuma myotis is closely associated with open water. Most commonly found in desert areas, this bat typically forages just above the surface of streams and ponds. Easterla (1973) examined the stomach contents of 14 *M. yumanensis* from Big Bend National Park, and found the following food items: moths (39.5%), froghoppers and leafhoppers (1.4%), June beetles (4.6%), ground beetles (2.5%), midges (12.9%), muscid flies (2.5%), caddisflies (2.1%), crane flies (1.4%), and unidentified insects (33.0%).

Parturition is believed to occur from late May to early June, with the female giving birth to one young per year. Large nursery colonies may form in buildings, caves, mine tunnels, and under bridges at this time. As with many other bats, the males take no part in care of the young and are not usually found near nursery colonies; instead they usually scatter and lead somewhat solitary lifestyles. Nursery colonies are very sensitive and quickly abandoned if disturbed. The nur-

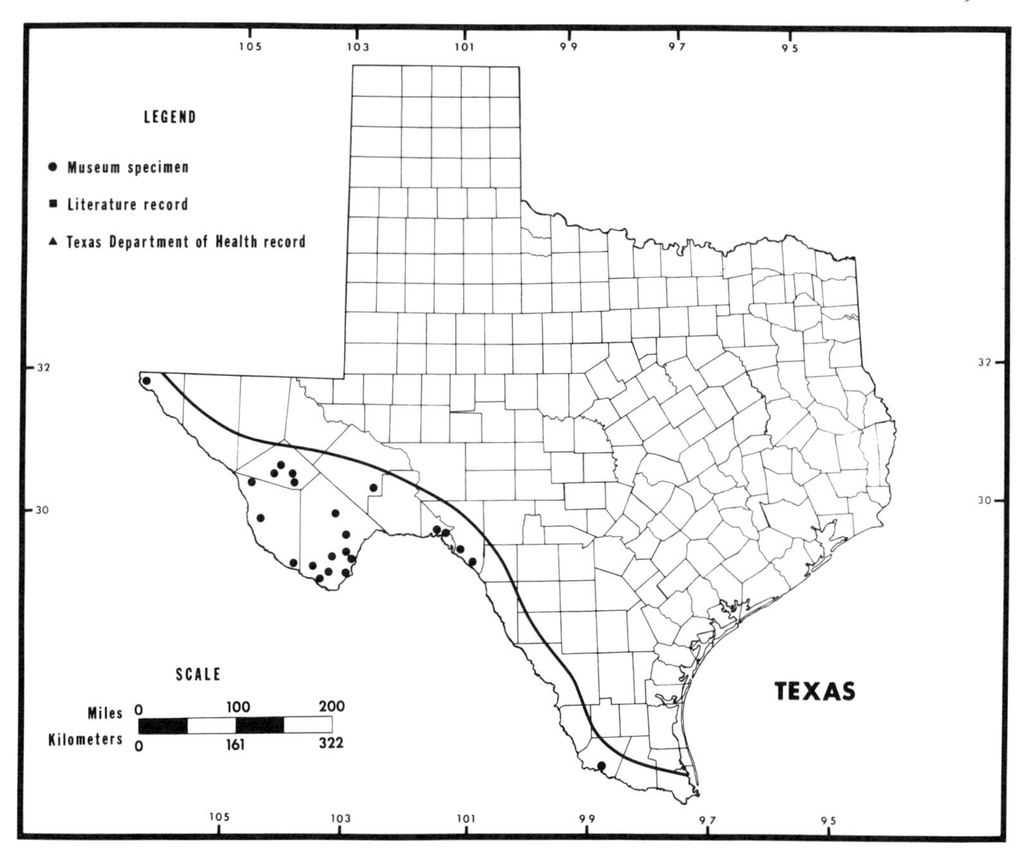

Map 3. Distribution of the Yuma myotis, *Myotis yumanensis yumanensis*.

sery colonies are eventually vacated in the fall, although the migrational destination of the bats is unknown. A few individuals may winter in Texas.

Specimens examined (114). *El Paso Co.:* El Paso, 1 (UTEP). *Pecos Co.:* 14.6 mi N, 19.2 mi E Marathon, 4,500 ft., 1 (TTU). *Jeff Davis Co.:* 3 mi E jct. hwys. 166 and 118, 1 (TTU); Sawtooth Mt., 3 (TTU); Espy tank, 0.5 mi SE Madera Canyon Campground, 1 (TTU); 2 mi S Ft. Davis, 1 (UMMZ); 8 mi S Ft. Davis, 1 (TTU). *Presidio Co.:* 11 mi W Valentine, 3 (TNHC); ZH Canyon, Sierra Vieja, 1 (TTU); Chinati Mts., 1 (TTU); 9 mi SW Valentine, 1 (CCSU); 3 mi NE Porvenir, 1 (FMNH); 30 mi SE Redford, 5 (MWSU). *Brewster Co.:* Marathon, 1 (USNM/FWS); 38 mi S, 14 mi E Marathon, 1 (UIMNH); BGWMA, 3 (DMNHT); 27 mi N Government Springs, 2 (AMNH); 4.5 mi SE Maravillas Creek on the Rio Grande, 1 (DMNHT); Terlingua ghost town, 4 (MWSU); 2 mi E Terlingua, 1 (TTU); Kennedy Ranch, 5 (DMNHT); Boot Springs, BBNP, 1 (UMMZ); Rio Grande Village waterhole, BBNP, 1 (BBNHA); Hot Springs Motel, BBNP, 1 (BBNHA); Kibee Springs, BBNP, 5,700 ft., 1 (FMNH); mouth Santa Elena Canyon, BBNP, 1 (MVZ); 1 mi SW Boquillas, 1,850 ft., BBNP, 2 (MVZ); BBNP, no specific locality, 2 (TTU). *Val Verde Co.:* 12 mi W Comstock, 2 (TTU); 10 mi W Comstock, 21 (TCWC); Comstock railroad tunnel, 12 mi W, 3 mi S Comstock, 3 (TTU); Mile Long Cave, 0.75 mi

E Langtry, 1 (TTU); 11 mi ESE Langtry, 18 (MWSU); jct. Pecos and Rio Grande rivers, 2 (USNM/FWS); Shumla, Old S.P. Tunnel, 5 (LACM); Shumla, Rio Grande River, 3 (LACM); 8 mi W Del Rio, 9 (USNM/FWS). *Starr Co.*: Rio Grande, 2 (TNHC).

References. 2, 3, 4, 7, 10, 12, 15, 21, 24, 29, 31, 33, 76, 84, 92, 99, 102, 105, 167, 181, 185, 188, 191, 192, 198, 220, 245, 256, 257, 265, 288, 306, 342, 414.

Myotis austroriparius (Rhoades, 1897)
Southeastern Myotis

Description. This is a small bat (forearm=36–41 mm, weight=5–7 g), with thick, woolly fur and no keel on the calcar. Long toe hairs extend beyond the tips of the claws. Two color phases are found in Texas. The most common bats are a dull gray to gray-brown color but occasionally bright, orange-brown individuals are encountered. Underparts are light brown to white.

The southeastern myotis is easily confused with two other species of *Myotis* in Texas—*M. septentrionalis*, which has relatively longer ears, and *M. lucifugus*, which may be distinguished by its glossy, dark brown fur (table 8). Average external measurements of *M. austroriparius* are: total length, 87 mm; tail, 34 mm; hind foot, 9 mm; ear, 14 mm.

Distribution. The southeastern myotis is known primarily from the Pineywoods of East Texas, although a few records have been reported from the Gulf Prairies and Marshes. It reaches the westernmost part of its North American range in these areas of Texas and is relatively uncommon here. Collecting records are intermittent throughout the year, suggesting that this bat is a year-round resident of the state.

Subspecies. This bat is monotypic and subspecies are not recognized, as indicated by the most recent taxonomic revision of the species (LaVal, 1970).

Life History. The southeastern myotis is another *Myotis* usually associated with water, where it feeds by skimming just above the water surface. Over most of its range in North America, this bat hibernates in a variety of winter roosting sites, particularly caves, but also culverts, bridges, buildings, and hollow trees—any place near water and protected from the vicissitudes of winter. In Texas, this bat is apparently active throughout the year and inhabits roosting sites other than caves, which are rare in east Texas. In contrast to summer maternity colonies, winter groups are small and scattered.

The southeastern myotis is unique among myotises in generally giving birth to twin offspring; other species usually have only one young per year. The breeding period is unknown, although in northern Florida they are believed to mate in both the spring and fall, with parturition occurring from late April to late May (Barbour and Davis, 1969). Maternity colonies can become quite large at this time, with the bats forming dense clusters of as many as ninety thousand indi-

Myotis austroriparius. Southeastern Myotis. (Merlin D. Tuttle, Bat Conservation International)

viduals. Reportedly, only caves containing standing water are used and males are occasionally present. After the young mature, males may join these colonies in increasing numbers, but they leave again by October when the populous nursery colonies begin to disperse.

Young bats weigh slightly over one gram at birth, but they grow quickly and are able to fly at five to six weeks. Both sexes reach sexual maturity before one year of age.

Because they are active for most of the year, it has been suggested that the higher fecundity of these bats may be in response to their higher mortality rate compared to other northern species which hibernate and are inactive a large part of the year (Rice, 1957). Both rat and corn snakes are known to frequent caves used by this species and may prey heavily upon these bats. Opossums and owls have also been reported to prey upon them, and cockroaches are known to feed upon young bats fallen to cave floors.

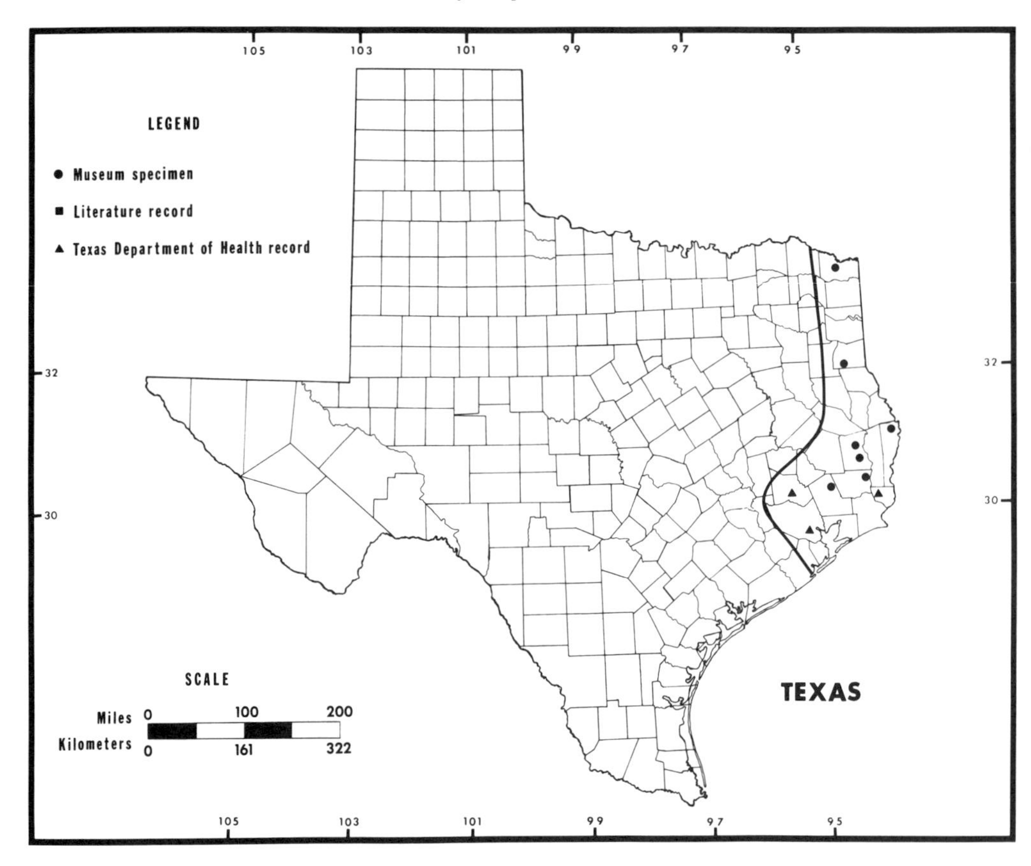

Map 4. Distribution of the southeastern myotis, *Myotis austroriparius.*

Specific food items are unknown, although small moths, midges, mosquitoes, and flies are probably of importance. Four specimens of this bat reported to the Texas Department of Health tested negative for rabies.

Specimens examined (23). *Bowie Co.*: New Boston, 1 (TTU). *Panola Co.*: 8 mi SW Gary, 1 (TCWC). *Newton Co.*: 12 mi N Burkeville, 6 (TCWC); 11.5 mi N Burkeville, 2 (TCWC); 8.5 mi N Burkeville, 2 (TCWC). *Tyler Co.*: 1.1 mi S, 1 mi W Town Bluff, BTNP, 3 (TCWC); 2 mi S, 1.5 mi W Town Bluff, BTNP, 1 (TCWC); 3.6 mi S, 2.9 mi W Town Bluff, BTNP, 1 (TCWC); 0.6 mi N, 0.7 mi W Spurger, BTNP, 2 (TCWC). *Hardin Co.*: 9 mi N, 1.3 mi E Silsbee, BTNP, 1 (TCWC); 10.9 mi N, 2.3 mi E Silsbee, BTNP, 1 (TCWC); 11.2 mi N, 2.3 mi E Silsbee, BTNP, 1 (TCWC). *Liberty Co.*: 2.5 mi N, 3.8 mi E Moss Hill, BTNP, 1 (TCWC).

References. 4, 7, 12, 13, 20, 21, 25, 27, 48, 84, 192, 265, 278, 300, 320, 352, 363.

Myotis velifer (J. A. Allen, 1890)
Cave Myotis

Description. This is the largest (forearm = 37–47 mm) species of the genus *Myotis* in Texas, with adults averaging about 15 g in weight. Males are usually slightly

Myotis velifer. Cave Myotis. (Merlin D. Tuttle, Bat Conservation International)

smaller than females. In the west, the cave myotis is sympatric with several other species of *Myotis* but its larger size; dark, dull brown coloration; and lack of a keeled calcar serve to distinguish it from the others (table 8). Average external measurements are: total length, 100 mm; tail, 42 mm; hind foot, 10 mm; ear, 15 mm.

Distribution. The cave myotis is a year-round resident of Texas, although it exhibits a varied seasonal distribution in the western two-thirds of the state. Summer months (March 16 to October 31) find it occupying the High Plains, Rolling Plains, Trans-Pecos, Edwards Plateau, and South Texas Plains, whereas during winter (November 1 to March 15) the species is apparently restricted to the central and north-central parts of the state. No winter records exist from the Trans-Pecos or Rio Grande Valley.

Seasonal variation in distribution of the sexes is not evident on the High Plains and Rolling Plains, but significantly more males than females are known from the Edwards Plateau during winter. In summer, males and females appear equally distributed across the Trans-Pecos, Edwards Plateau, and South Texas Plains.

Subspecies. Texas specimens are referrable to two subspecies according to the most recent taxonomic revision of the species in Texas (Dalquest and Stangl, 1984a). M. *v. incautus* (J. A. Allen, 1896) is found in the Trans-Pecos, Edwards Plateau, and South Texas Plains regions and M. *v. magnamolaris* Choate and Hall, 1967,

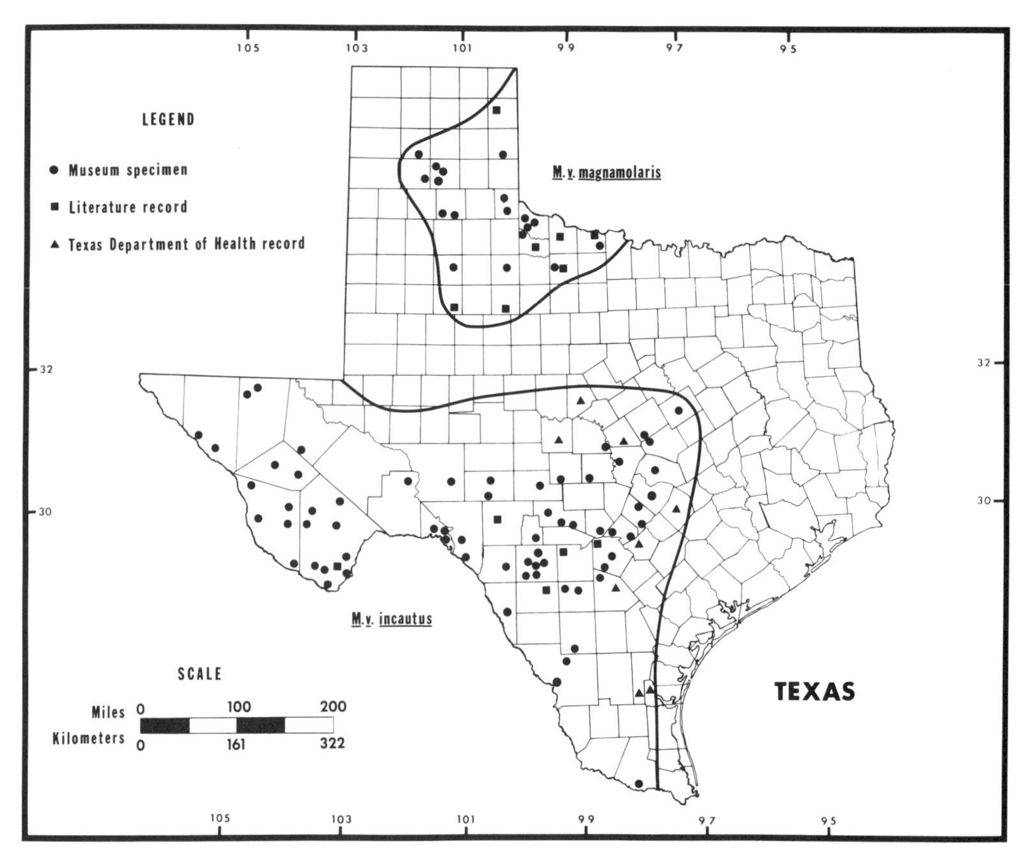

Map 5. Distribution of the two subspecies of the cave myotis, *Myotis velifer.*

occupies the High Plains and Rolling Plains of the Texas Panhandle. *M. v. magnamolaris* is paler in coloration, averages 5 mm longer in total length, and has slightly greater skull measurements than *M. v. incautus* (Choate and Hall, 1967).

Life History. The cave myotis is the most abundant myotis of the Edwards Plateau region and is known to hibernate in central Texas caves during winter. The species also hibernates in the gypsum caves found throughout the Panhandle and rarely makes major movements during winter, although short flights between hibernacula occasionally occur. The bats usually roost in caves and tunnels, typically in clusters numbering into the thousands, but during migration they have been found in rock fissures, carports, occasionally in attics and old buildings, and even in abandoned cliff swallow nests (Buchanan, 1958; Hayward, 1970). Several other bats, including Townsend's big-eared bats (*Plecotus townsendii*), Brazilian free-tailed bats (*Tadarida brasiliensis*), big brown bats (*Eptesicus fuscus*), Yuma myotises (*Myotis yumanensis*), and ghost-faced bats (*Mormoops megalophylla*) often hibernate at the same sites as *Myotis velifer,* although the bats usually segregate by species, with different kinds inhabiting separate areas or rooms of the roosting site.

In summer, cave myotises typically leave their roosts 10–15 minutes after sunset to forage for food and to find water. The large size and stronger, more direct flight of the cave myotis, relative to other *Myotis* species, help to distinguish it from others feeding in the same area. Also, cave myotises generally fly approximately 4–12 m above ground while foraging and tend to fly close to vegetation (Fitch et al., 1981).

As with most other insectivorous bats, the cave myotis is an opportunistic feeder, taking a wide variety of insects depending upon what is available on a particular evening. A study of the species in Arizona (Hayward, 1970) revealed that small moths make up the largest part of their diet, although weevils, antlions, and several species of small beetles are also taken. Banding studies (Tinkle and Patterson, 1965; Twente, 1955b) have shown that these bats do not always return to the same roosting sites on succeeding nights; in combination with their relatively powerful flying ability, this may explain why they are able to forage farther abroad than other species of *Myotis.*

Data on their reproductive habits are sparse. In Arizona embryo implantation is believed to occur by the first of May with parturition occurring in the last week of June, resulting in a gestation period of 45–55 days (Hayward, 1970). In Texas, females have been found carrying embryos as early as April 12, which suggests that individuals in our state may give birth earlier than in other areas. On the Edwards Plateau, lactating females have been captured frequently in May, supporting the idea that parturition may occur as early as the first two weeks of May in Texas.

One young per year is born to the female. Baby bats are "hung" in a nursery colony with other newborns, where they are nursed and protected by the adult females. Such nursery colonies can be quite large—up to 15,000 individuals—but generally contain between 1,000 and 5,000 bats. The young are capable of flight at about five weeks of age and, until that time, may be moved to a different location by the mothers if the nursery colony is disturbed.

Of 82 cave myotises reported to the Texas Health Department, only 2 tested positive for rabies.

Specimens examined. *Myotis velifer magnamolaris* (264). *Potter Co.*: Amarillo, 1 (KU). *Wheeler Co.*: Shamrock, 1 (TTU). *Randall Co.*: cave in Palo Duro Canyon, 3 (TCWC). *Armstrong Co.*: 12 mi SE Washburn, 3,000 ft., 16 (TCWC); 8 mi S Claude, 16 (TTU); 10 mi NE Wayside, east side Red River, 5 (KU). *Briscoe Co.*: Los Lingos Canyon, 10 mi WSW Quitaque, 3 (TTU); 9 mi N South Plains, hwy. 207, 1 (TTU). *Childress Co.*: 16 mi N Childress, Lost Cave, 10 (TTU); 1.6 mi N Childress, 10 (TTU); Panther Cave, 22 mi SE Childress, 6 (TTU). *Hardeman Co.*: 3 mi S Lazare, 4 (UIMNH); 3 mi SE Lazare, 24 (18 MWSU, 6 TTU); Acme, 16 (14 MWSU, 2 KU); Acme Mine Shaft, 4 (TTU); 9 mi W Quanah, 1 (MWSU); Quanah, 1 (MWSU); 3 mi SW Quanah, 1 (MWSU); 4 mi SW Quanah, 1 (MWSU); Walkup/Lazare Cave, 9 mi SW Quanah, 3 (TTU); 10 mi SW Quanah, 2 (MWSU); 11 mi WSW Quanah, 45 (MWSU); 11 mi SW Quanah, Beasley Cave, 7 (MWSU); Gregory Cave, 19 (TTU). *Foard Co.*: 24 mi W Quanah, 1 (MWSU); 18 mi W Quanah, 1 (MWSU); 15 mi W Quanah, 2 (MWSU); 14 mi W Quanah, 1 (MWSU); 12 mi SW Lazare, 1 (MWSU); 12 mi SW Quanah, 21 (MWSU); 13 mi WSW Quanah, 2 (MWSU); 12 mi SSW Quanah, 3 (MWSU); 13.5 mi S Quanah, Pease River, 12 (MWSU); 15 mi

SW Quanah, 1 (MWSU). *Wichita Co.*: 1 mi N Wichita Falls, 1 (MWSU). *Crosby Co.*: 4 mi E Crosbyton, 1 (TTU). *King Co.*: River Styx Cave, 10 mi NE Guthrie, 7 (TTU); 8 mi NE Guthrie, 1 (TTU); 6 mi E Guthrie, 6 (TCWC); 10 mi SE Guthrie, 2 (TTU). *Baylor Co.*: 5 mi NW Redsprings, 1 (MWSU).

Additional records: *Hemphill Co.*: 5 mi E Canadian (Milstead and Tinkle, 1959). *Wilbarger Co.*: Vernon (Dalquest, 1968). *Foard Co.*: Crowell (Milstead and Tinkle, 1959). *Wichita Co.*: 1.6 km N Burkburnett (Cokendolpher et al., 1979). *Baylor Co.*: 4 mi E Seymour (Baccus, 1971). *Garza Co.*: Justiceburg (Milstead and Tinkle, 1959). *Stonewall Co.*: 7.4 mi NW, 6.3 mi W Hamlin (Hayward, 1970).

Myotis velifer incautus (818). *Hudspeth Co.*: 2 mi W Fort Hancock, 3 (KU); 12 mi S Sierra Blanca, 1 (MWSU). *Culberson Co.*: McKittrick Canyon, GMNP, 4 (TCWC); 4 mi E Pine Springs, GMNP 4,500 ft., 1 (TCWC). *Reeves Co.*: Toyahvale, 7 (UMMZ); near Toyahvale and Phantom Lake, 38 (UMMZ). *McLennan Co.*: Science Hall, Baylor Univ., Waco, 1 (SM). *Coryell Co.*: Fort Hood, 1 (FMNH). *San Saba Co.*: 5.0 mi SSE Bend, Gorman Falls Fish Camp, 35 (TCWC). *Bell Co.*: Fort Hood, 1 (KU). *Jeff Davis Co.*: 3 mi E jct. hwys. 166 and 118, 3 (TTU); 8 mi N Ft. Davis, 1 (TTU); 8 mi NE Ft. Davis, 1 (TCWC); Sawtooth Mt., Davis Mts., 1 (SRSU); 3 mi E DMSP, 2 (TTU); mouth Madera Canyon, Davis Mts., 4,400 ft., 1 (TCWC); 3.5 mi NE Ft. Davis, 3 (TTU). *Crockett Co.*: 23 mi SE Ozona, 1 (TCWC); 23.5 mi SE Ozona, Friend Ranch, 9 (TCWC); 24 mi SE Ozona, Friend Ranch, 1 (TCWC); 25 mi SSE Ozona, Bagget Ranch, 5 (TCWC). *Burnet Co.*: 21 mi SW Lampasas, 2 (USNM/FWS). *Mason Co.*: James River Bat Cave, 10 (TTU). *Llano Co.*: Enchanted Park, 1 (TTU). *Williamson Co.*: 3 mi N Georgetown, 1 (TNHC); 3 mi W Georgetown, 4 Mile Cave, 3 (TNHC); 1.25 mi SW Georgetown, Steam Cave, 4 (UIMNH); 2 mi SW Georgetown, 1 (TNHC); 3 mi SW Georgetown, 1 (TNHC); 2.5 mi NW McNeal, Merril Cave, 9 (TNHC). *Terrell Co.*: 21 mi S Sheffield, 1 (TNHC). *Sutton Co.*: Fulton Cave, 22 (TTU); 14.3 mi E Sonora, 1 (MVZ); 31 mi SE Sonora, 7 (USNM/FWS). *Kimble Co.*: 2.2 mi S, 8 mi E Junction, 12 (TTU). *Presidio Co.*: ZH Canyon, Sierra Vieja, 4 (TTU); San Esteban Lake on Casa Piedra Rd., 1 (SRSU); 8 mi S Marfa, 5 (SRSU); 10 mi S Marfa 4,300 ft., 80 (51 LACM, 29 TCWC); 10 mi S Marfa, 5 mi S San Esteban Dam, 3 (TCWC); 12 mi SSE Marfa, 4,600 ft., 22 (TCWC); Dead (Wild) Horse Canyon, Chinati Mts., 1 (SRSU); Chinati Mts., 4 (TTU); 37 mi S Marfa, 1 (TCWC); 12 mi W Lajitas, 1 (SRSU). *Brewster Co.*: 21.2 mi N, 1 mi E Marathon, 1 (TTU); 13.2 mi N, 2.6 mi E Marathon, 5,200 ft., 1 (TTU); Woodward Campground, 4 (SRSU); Calamity Creek bridge, 22 mi S Alpine, 1 (SRSU); 25 mi S Alpine, 1 (TCWC); 16 mi S Marathon 3,900 ft., 1 (TCWC); BGWMA, 3 (2 TCWC, 1 DMNHT); 0.25 mi N Terlingua, 1 (SRSU); Maverick, BBNP, 1 (SRSU); Big 38 Mine, 3 mi W Terlingua, 4 (MWSU); Rio Grande Village waterhole, BBNP, 1 (BBNHA); Big Bend of the Rio Grande, 2,000 ft., 6 mi E Johnson's Ranch House, BBNP, 4 (TCWC). *Travis Co.*: 1 mi NW Austin, 1 (TNHC); Austin, 1 (TNHC); Austin, cave on Balcones Trail, 4 (TNHC); Austin, 5 mi E UT Campus, 1 (TNHC); Austin, 7 mi SW UT Tower, 2 (TNHC). *Blanco Co.*: no specific locality, 5 (FMNH). *Hays Co.*: 10 mi NW Buda, 17 (FMNH); 2 mi S, 5 mi E Dripping Springs, 1 (KU); San Marcos, 1 (FMNH); Camp Ben McCulloch, 1 (MSU). *Val Verde Co.*: 44 mi N, 6 mi W Del Rio, 1 (TTU); 3 mi NNE Langtry, 15 (MWSU); 2 mi NW Langtry, Fisher's Fissure, 9 (TTU); Pump Canyon, Langtry, 2 (USNM/FWS); Langtry, 1 (CCSU); Mice Canyon, 0.75 mi E Langtry, 3 (TTU); 12 mi N Comstock, 2 (MWSU); Fawcett Cave, 3.6 mi N Del Rio, 6 (TTU); 11 mi ESE Langtry, 4 (MWSU); 1 mi W Pecos River on Rio Grande, 5 (SRSU); mouth of Pecos River, 6 (1 KU, 1 TCWC, 4 USNM/FWS); Shumla, 18 (LACM); 10 mi W Comstock, 1,300 ft., 1 (TCWC); Comstock, 2 (TTU); 5.4 mi E Comstock, 3 (CCSU); 12 mi W, 3 mi S Comstock, 19 (TTU); Dolan Snake Springs, 12 mi W, 5 mi S Comstock, 3 (TTU); cave, 4 mi N Del Rio, 1 (TCWC); east Painted Cave, 1 (USNM/FWS). *Kerr Co.*: 13 mi W Hunt, Kerr Wildlife Mgt. Area, 11 (TCWC); 8 mi SW Kerrville, 4 (TCWC); 15 mi SW Kerrville, 2,000 ft., 1 (MVZ); 11 mi N Medina, 3 (TCWC); Camp Verde, Verde Creek, 2 (CCSU); Japonica, 7 (USNM/FWS). *Kendall Co.*: Boerne, 2 (FMNH); Cascade Caverns, 4.6 mi SE Boerne, 1 (TCWC); Cascade Caverns, 4.8 mi SE Boerne, 1 (MVZ). *Real Co.*: Leaky, 2 (THNC). *Comal Co.*: 5 mi NW Bracken Cave, 13 (LSUMZ); 5 mi N Bracken Cave, 6 (LSUMZ); Bracken Cave, 10 (1 AMNH, 1 KU, 2 LACM, 6 TCWC); *Bexar Co.*: San Antonio, 6 (4 AMNH, 2 USNM/FWS); Somerset, 1 (KU); W Hightower, 1 (TNHC). *Kinney Co.*: 14.1 mi E Brackettville, hwy. 90, 12 (TCWC). *Uvalde Co.*: Concan, 27 (LACM); 25 mi NW Uvalde, off hwy. 55, Cal Newton Ranch, 2 (TCWC); Frio Cave, 164 (LACM); 16 mi NW Sabinal, 3 (MVZ); 20 mi N Uvalde, 3 (USNM/FWS); 2.5 mi N Uvalde, 1 (CCSU); 3.9 mi E Cline, 1 (TNHC); 3.2 mi E Cline on hwy. 90, 5 (1 TCWC, 4 TNHC); 14.3 mi W Uvalde, 6 (TNHC); 11.9 mi W Uvalde, 1 (TNHC); 0.25 mi S, 1 mi W Uvalde, 2 (TCWC). *Atascosa Co.*: 7 mi SW Somerset, 1 (TNHC). *Maverick Co.*: 23 mi E Eagle Pass, hwy. 277, 1 (TAIU). *Frio Co.*: 3.5 mi S Moore, 10 (TNHC); Frio Town, 1 (TCWC); 6.2 mi SW Moore, IH 35 (Calvert 301), 8 (TNHC); 7.1 mi SW Moore, 1 (TNHC). *La Salle Co.*: 8 mi E Encinal, 2 (TCWC). *Webb Co.*: 6 mi S Encinal, San Ramon Ranch, 11 (TCWC); 4 mi N Laredo, 5 (KU). *Hidalgo Co.*: Santa Ana Refuge, 8.5 mi S Alamo, 2 (USNM/FWS).

Additional records: *Dallas Co.*: no specific locality (Davis, 1974). *Presidio Co.*: Marfa (Constantine, 1957). *Brewster Co.*: Giant Dagger Yucca Flats (Easterla, 1968). *Hays Co.*: Buda (Constantine, 1957). *Kerr Co.*: Kerrville (Constantine, 1957). *Edwards Co.*: Dunbar Cave, 21 mi W Rocksprings (Selander and Baker, 1957). *Bexar Co.*: Haby Cave (Eads et al., 1956); San Antonio (Wiseman et al., 1962). *Medina Co.*: Valdina Farms Sinkhole (Raun and Baker, 1959). *Zavalla Co.*: hwy. 57, 2 km W Batesville (Jackson et al., 1982). *La Salle Co.*: Encinal (Constantine, 1957).

References. 1, 2, 3, 4, 6, 7, 10, 12, 15, 21, 24, 25, 29, 33, 41, 63, 73, 76, 84, 96, 97, 98, 102, 107, 110, 115, 116, 123, 127, 129, 131, 145, 151, 153, 154, 155, 168, 170, 171, 181, 188, 191, 192, 198, 201, 202, 223, 226, 234, 245, 251, 253, 255, 256, 261, 262, 265, 267, 272, 274, 276, 279, 288, 293, 302, 304, 306, 309, 332, 337, 340, 342, 356, 358, 362, 366, 387, 391, 397, 398, 400, 401, 402, 406, 414.

Myotis septentrionalis (Trouessart, 1897)
Northern Myotis

Description. This is a small to medium-sized (forearm = 32–39 mm) bat with dull, gray-brown pelage. Other characteristics include relatively long ears, compared

Myotis septentrionalis. Northern Myotis. (Merlin D. Tuttle, Bat Conservation International)

with other species of *Myotis* in Texas, and an unkeeled calcar (table 8). As with all myotis species, this bat also has a narrow and sharp-pointed tragus. Average external measurements are: total length, 78 mm; tail, 26 mm; hind foot, 9 mm; ear, 13 mm.

Distribution. M. *septentrionalis* is widely distributed over eastern and northern North America, but is known in Texas on the basis of a single specimen collected at Winter Haven, Dimmit County, in South Texas. This specimen (consisting of a skin and skull) was sent by S. E. Jones to the Division of Insects, U.S. Bureau of Entomology, United States National Museum, on August 19, 1942. The nearest known locality of this species is from Arkansas, which is over 800 km to the northeast of Dimmit County. It is doubtful that resident populations of this species occur in Texas.

Subspecies. M. *septentrionalis* (Trouessart, 1897) is a monotypic species and subspecies are not recognized.

Life History. The northern myotis commonly hibernates in caves and mine tunnels of eastern Canada and in the United States from Vermont to Nebraska (Barbour and Davis, 1969). M. *septentrionalis* seems much more solitary in its habits than other myotises, and it is generally found singly or in small groups containing up to a hundred individuals. Tuttle and Heaney (1974) located twelve roosts in South Dakota, of which only two contained more than a single adult or an adult and its offspring. This bat is known to share hibernacula with the little brown myotis (*Myotis lucifugus*), big brown bat (*Eptesicus fuscus*), and eastern pipistrelle (*Pipistrellus subflavus*) and generally chooses cooler sites with high humidity for hibernation (Barbour and Davis, 1969). *M. septentrionalis* is also known to roost in hollow trees and rock crevices, and in summer has been found behind tree bark and in buildings.

Nothing is known about its food habits, although it commonly forages along forest edges, over forest clearings, and occasionally over ponds. Likewise, the reproductive habits are poorly known, except that small nursery colonies seem the rule. Tuttle and Heaney (1974) found a maternity roost in South Dakota that contained four lactating females nursing five nonvolant juveniles, which suggests that twinning may possibly occur in this species.

Remarks. Although it was previously considered a subspecies of *Myotis keenii* (*M. k. septentrionalis*), van Zyll de Jong (1979) elevated *M. septentrionalis* to full species status based upon cranial, dental, and external characters. This taxonomic rearrangement created two monotypic species, *M. keenii* of the northwestern United States and Canada, and the paler *M. septentrionalis* of eastern North America.

Specimens examined (1). *Dimmit Co.*: Winter Haven, 1 (USNM/FWS).

References. 1, 4, 12, 22, 27, 31, 33, 40, 84, 123, 142, 188, 198, 243, 262, 265, 404.

Myotis thysanodes G. S. Miller, 1897
Fringed Myotis

Description. This is the most easily recognized species of *Myotis* occurring in Texas. A conspicuous fringe of short hairs lines the free edge of the interfemoral membrane, from which the species gets its common name. Pelage coloration is buff brown above and dull white below and, compared with other myotis species, the ears are long and the feet large (table 8). Average external measurements are: total length, 85 mm; tail, 37 mm; hind foot, 10 mm; ear, 17 mm.

Distribution. This species is distributed throughout the western United States and is known in Texas from mountainous areas of the Trans-Pecos during summer months. Two specimens were captured in west-central Texas (Crosby County; Jones et al., 1987), but these may have been seasonal migrants. The fringed myotis has been collected in habitats ranging from mountainous pine, oak, and piñon-juniper to desert scrub, but seems to prefer grassland areas at intermediate elevations. Barbour and Davis (1969) report capturing a specimen in spruce-fir forest at 2,850 m in New Mexico, but these authors note that *M. thysanodes* is typically encountered at lower elevations of 1,200–2,150 m. No winter records are available for this bat in Texas and its winter habits remain unknown.

Subspecies. Texas specimens are referrable to the subspecies *M. t. thysanodes* G. S. Miller, 1897 (O'Farrell and Studier, 1980).

Myotis thysanodes. Fringed Myotis. (John L. Tveten)

Life History. *M. thysanodes* is a migratory species that arrives in the Trans-Pecos by May, at which time it begins forming nursery colonies. It is a colonial bat and maternity roosts may contain several hundred individuals. It has been found inhabiting caves, mine tunnels, rock crevices, and old buildings.

Its reproductive biology in Texas is poorly known. A study conducted in New Mexico (O'Farrell and Studier, 1973) revealed that parturition occurred within a two-week period—from the end of June to early July—after gestation of 50 to 60 days. Young bats are able to fly at 16 to 17 days and, until that time, are guarded at night roosts by "nurse bats"—adult females that remain with the nursery colony to protect and watch over the baby bats while the rest of the adults are out foraging. Males are usually absent from maternity roosts, and these colonies begin to disperse by October. As mentioned above, winter locales and habits of the fringed myotis remain a mystery.

Black (1974) studied the food habits of this species in New Mexico and found that it mainly ate beetles. His observations indicate relatively slow, highly maneuverable flight, with foraging concentrated close to the vegetative canopy.

Specimens examined (89). *El Paso Co.*: 20 mi E El Paso, 4,500 ft., 1 (AMNH). *Hudspeth Co.*: Eagle Bluff, Eagle Mts., 6,500 ft., 2 (UTEP). *Culberson Co.*: McKittrick Canyon, GMNP, 12 (TCWC); Lost Peak, GMNP, 2 (TTU); Manzanita Spring, GMNP, 1 (TTU); Smith Spring, GMNP, 1 (TTU); The Bowl, GMNP, 4 (TTU). *Jeff Davis Co.*: 8 mi W jct. hwys. 166 and 118, 2 (SRSU); 3 mi E jct. hwys. 166 and 118, 1 (TTU); Rockpile Park on hwy. 118, 5 (SRSU); 8 mi S jct. hwys. 166 and 118, 3 (TTU); Davis Mts., 2.3 mi W hwy. 166 on hwy. to Valentine, 1 (ASVRC);

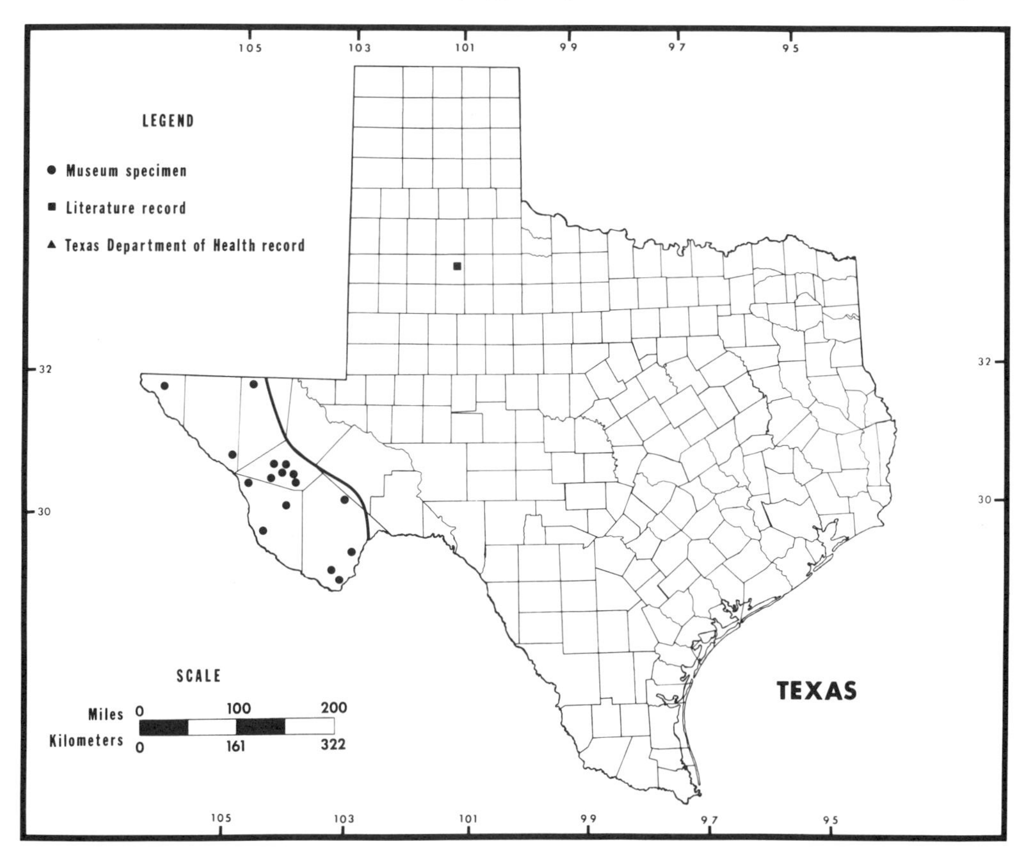

Map 6. Distribution of the fringed myotis, *Myotis thysanodes thysanodes.*

3 mi E jct. hwys. 166 and 118, 1 (TTU); Sawtooth Mt., 3 (TTU); DMSP, 2 (TTU); 8 mi S Ft. Davis, 1 (TTU). *Presidio Co.:* Sierra Vieja, 9 mi SW Valentine, 5 (TTU); ZH Canyon, Sierra Vieja, 1 (TTU); 10 mi S Marfa, 4,300 ft., 2 (TCWC); Chinati Mts., Pinto Canyon, 10 (5 TTU, 1 UIMNH, 4 TCWC); Upper Pinto Canyon, Chinati Mts., 1 (SRSU); Chinati Peak, 7,700 ft., 3 (TCWC); Upper Wild Horse Canyon, Chinati Mts., 1 (SRSU). *Brewster Co.:* 13.2 mi N, 2.6 mi E Marathon, 5,200 ft., 1 (TTU); BGWMA, 15 (13 DMNHT, 2 LACM); Pine Canyon, BBNP, 2 (BBNHA); Emory Peak, BBNP, 1 (FMNH); SE slope Mariscal Mt., BBNP, 2,800 ft., 5 (MVZ).

Additional records: *Crosby Co.:* 4 mi E Crosbyton (Jones et al., 1987).

References. 2, 4, 7, 10, 12, 15, 21, 24, 29, 31, 33, 52, 84, 92, 99, 105, 142, 167, 181, 188, 191, 192, 198, 201, 245, 256, 257, 264, 265, 267, 270, 279, 306, 314, 316, 342, 414.

Myotis volans (H. Allen, 1866)
Long-legged Myotis

Description. The long-legged myotis is a rather large *Myotis* (forearm=35–41 mm), which may be distinguished by a combination of features including a keeled cal-

Myotis volans. Long-legged Myotis. (Roger W. Barbour)

car, relatively long tail, short ears, and large feet (table 8). Most importantly, the underside of the wing membrane is lightly furred to an imaginary line connecting the elbow with the knee. Pelage coloration is russet to dark brown and the ears are dark, almost black. Average external measurements are: total length, 95 mm; tail, 43 mm; hind foot, 8 mm; ear, 11 mm.

Distribution. Although distributed throughout western North America, the long-legged myotis apparently is a relatively rare bat in Texas. It is known primarily from the Trans-Pecos region, where it seems to prefer high, open woods and mountainous terrain in the Guadalupe, Davis, Chinati, Chisos, and Sierra Vieja mountain ranges. A single, enigmatic specimen has been taken from Knox County in the Rolling Plains region. Originally reported as *M. lucifugus* (Baker, 1964; Mollhagen and Baker, 1972), this bat was found in a region of much local relief, in keeping with this species' habitat preferences. This was probably a wandering individual, and resident populations are not believed to inhabit the Rolling Plains region.

No winter records are available for this species in Texas, and it is most likely only a summer resident of the state.

Subspecies. Texas specimens are referrable to the subspecies *M. v. interior* G. S. Miller, 1914.

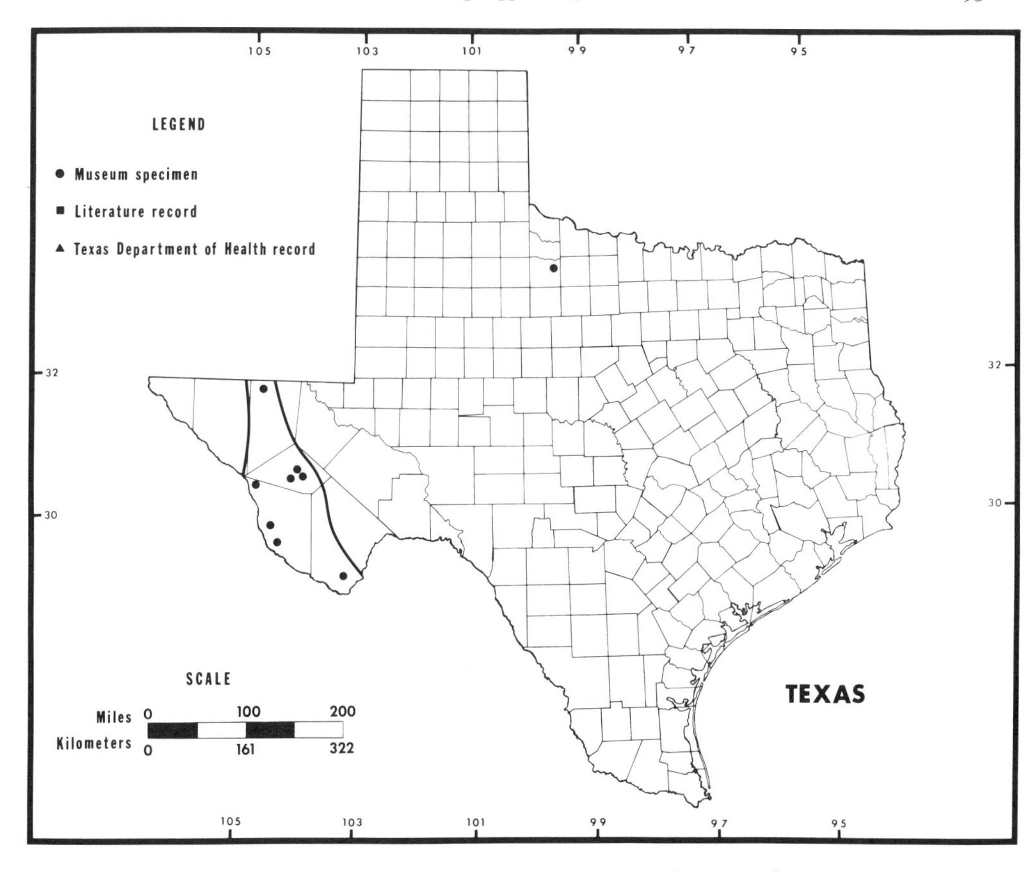

Map 7. Distribution of the long-legged myotis, *Myotis volans interior.*

Life History. Winter range and habits of the long-legged myotis are poorly understood. Nursery colonies, which may contain several hundred individuals, form in summer in places such as buildings, cliff crevices, and hollow trees. Reportedly, these bats do not use caves as day roosts, although they may use such sites at night (Barbour and Davis, 1969). These bats emerge early in the evening to forage over water and forest clearings. Evidence from New Mexico suggests that M. *volans* forages primarily in the open and feeds mainly on small moths (Black, 1974).

Only one young is born annually to the females, and birth is thought to occur earlier in the year than for most other bats. The timing of parturition in Texas is probably in June or early July.

Specimens Examined (30). *Knox Co.:* 3 mi N Vera, 1 (MSU). *Culberson Co.:* Manzanita Spring, GMNP, 1 (TTU); The Bowl, GMNP, 2 (TTU); McKittrick Canyon, GMNP, 2 (TCWC). *Jeff Davis Co.:* 0.5 mi SE Madera Canyon, Davis Mts., 1 (TTU); DMSP, 2 (TTU); Limpia Canyon, DMSP, 2 (TTU); 5 mi E Mt. Livermore, 1 (UMMZ). *Presidio Co.:* 11 mi W Valentine, 1 (TNHC); ZH Canyon, Sierra Vieja, 9 mi W Valentine, 2 (TTU); Chinati Mts., 5,000 ft., 1 (TCWC); Chinati Peak, 7,700 ft., 3 (TCWC); Chinati Mts., 5 (TTU); Pinto Canyon, Chinati Mts., 14 mi E Ruidosa, 2 (UIMNH); Pinto Canyon, Chinati Mts., 45 mi SW Marfa,

1 (TCWC). *Brewster Co.:* Boot Springs, BBNP, 1 (TCWC); Kibee Spring, Chisos Mts., 5,700 ft., BBNP, 1 (FMNH); Basin Sewage Lagoon, BBNP, 1 (BBNHA).

References. 2, 3, 4, 6, 7, 10, 12, 21, 24, 29, 31, 33, 57, 84, 92, 93, 99, 142, 181, 188, 191, 192, 198, 201, 245, 256, 257, 265, 279, 306, 311, 342, 414.

Myotis californicus (Audubon and Bachman, 1842)
California Myotis

Description. This is the smallest (forearm = 29–36 mm) species of *Myotis* in Texas. Coloration is light brown, the ears are black, and a dark facial mask extends across the eyes and rostrum from ear to ear. The calcar is keeled, and compared with other myotises, the feet are tiny and the tail and ears relatively long (table 8).

This bat is easily confused with the western small-footed myotis (*M. ciliolabrum*), from which it differs in having a sharply rising braincase, in contrast to the flattened skull of *M. ciliolabrum.* This cranial feature gives the California myotis a more prominent forehead than is evident in the western small-footed myotis (Bogan, 1974; van Zyll de Jong, 1984).

The western pipistrelle (*Pipistrellus hesperus*) also resembles this species, although it may be easily distinguished by its slightly smaller size and short, blunt tragus as opposed to the sharp-pointed tragus of *M. californicus.* Average external measurements are: total length, 79 mm; tail, 36 mm; hind foot, 6 mm; ear, 13 mm.

Distribution. The California myotis, which commonly occurs throughout the western United States and Mexico, is known in Texas exclusively from the Trans-Pecos, where it has been found in desert, grassland, and wooded habitats. Apparently, this is one of the few species that winters in the Trans-Pecos. Torpid individuals have been discovered in irrigation tunnels in Presidio County during December (Young and Scudday, 1975) and in the Franklin Mountains of El Paso County in January (Dooley, 1974).

Subspecies. Texas specimens are referrable to the subspecies *M. c. californicus* (Audubon and Bachman, 1842), as indicated by the most recent taxonomic revision of the species (Bogan, 1975).

Life History. The California myotis typically roosts in a variety of crevice-like places such as rock fissures or behind signs and loose tree bark, and it may frequently be found in buildings. This bat uses man-made structures as roosts more than any other species of *Myotis* (Schmidly, 1977). In winter, caves, mine tunnels, and buildings are used as hibernacula. Colonies are usually small, consisting of up to twenty-five individuals. These bats apparently do not form the compact clusters typical of many other species. Winter records are abundant from the southwestern United States, and this species is fairly active through the winter months. In summer, it is quite transient and will use any suitable and immediately available site for shelter.

Myotis californicus. California Myotis. (J. Scott Altenbach)

A single young is born from late May to early June. Maternity colonies are typically small and, as with many other species, growth is quite rapid. By mid-July, the young are of adult size and volant.

Data on feeding habits are lacking. Specific food items are unknown, but this bat appears to feed primarily on small moths and beetles that occur between, within, or below the vegetative canopy (Black, 1974).

Specimens examined (112). *El Paso Co.:* 4 mi NNW El Paso, 1 (KU); El Paso, 2 (1 USNM/FWS, 1 UTEP). *Culberson Co.:* 7 mi N Pine Springs, GMNP, 1 (TCWC); McKittrick Canyon, GMNP, 5,500 ft., 10 (TCWC); Smith Spring, GMNP, 1 (TTU); Signal Peak, GMNP, 1 (TTU). *Presidio Co.:* 11 mi W Valentine, 4 (TNHC); Sierra Vieja, 9 mi SW Valentine, 1 (TTU); ZH Canyon, Sierra Vieja, 2 (TTU); San Esteban Tunnel, 1 (SRSU); Chinati Mts., Pinto Canyon, 14 mi E Ruidosa, 2 (TTU); 3 mi NE Porvenir, 1 (FMNH); 5 mi SE Bandera Mesa, 14 (13 MWSU, 1 TCWC). *Brewster Co.:* 13.2 mi N, 2.6 mi E Marathon, 5,200 ft., 1 (TTU); Paisano, 1 (USNM/ FWS); 15 mi N Study Butte, 9 (MWSU); Musquiz Canyon, 3 (DMNHT); BGWMA, 46 (2 TTU, 7 TCWC, 37 DMNHT); 20 mi W LaLinda, 1 (SRSU); Burnham Ranch, Government Spring, 1 (AMNH); 3 mi S Government Spring, 1 (AMNH); Panther Junction, BBNP, 1 (BBNHA); Pine Canyon, BBNP, 2 (BBNHA), E base Burro Mesa, BBNP, 3,500 ft., 3 (MVZ); Santa Elena Campground, BBNP, 2 (TCWC).

Additional records: *Jeff Davis Co.:* no specific locality (Davis, 1974).

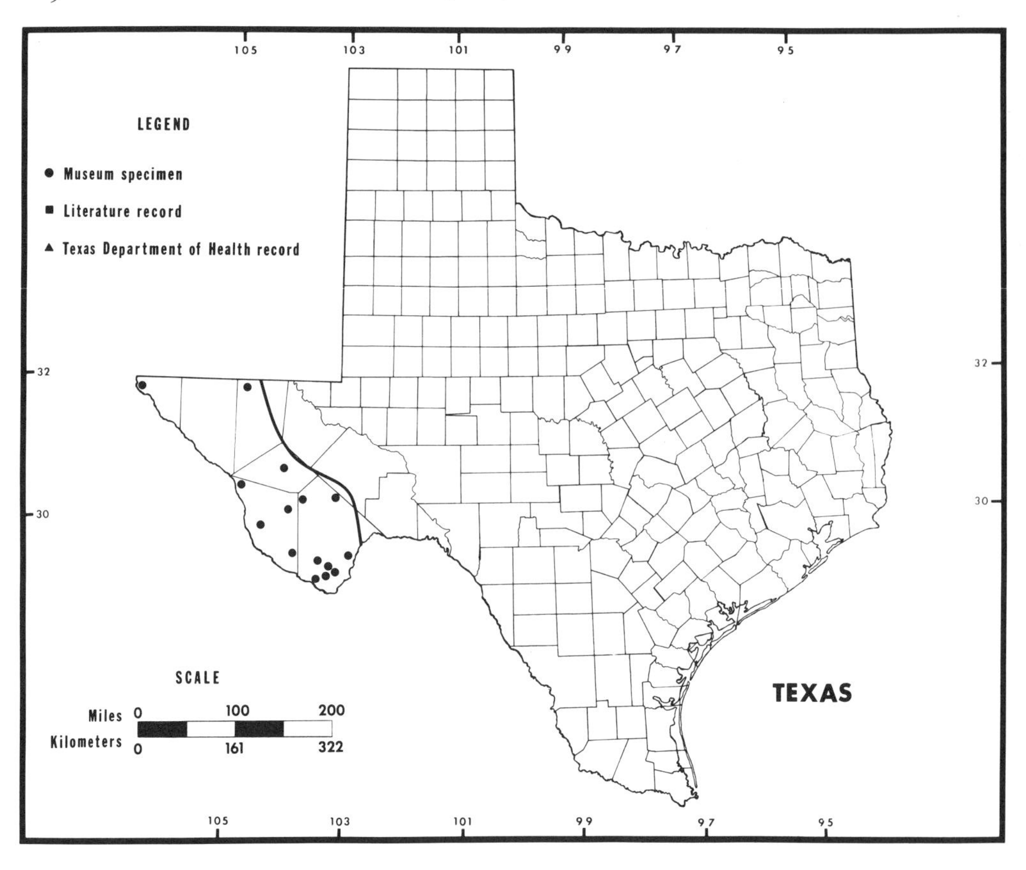

Map 8. Distribution of the California myotis, *Myotis californicus californicus.*

References. 1, 2, 3, 4, 7, 10, 12, 15, 21, 24, 29, 31, 33, 76, 84, 92, 99, 100, 101, 105, 155, 163, 166, 167, 181, 188, 191, 192, 198, 201, 245, 256, 257, 265, 279, 306, 358, 362, 427.

Myotis ciliolabrum (Merriam, 1886)
Western Small-footed Myotis

Description. This small *Myotis* (forearm=30–36 mm, weight=4–5 g) has buff brown pelage and black ears (table 8). The calcar is keeled and a dark facial mask is present across the rostrum and eyes from ear to ear. This bat is easily confused with the California myotis (*Myotis californicus*) and the western pipistrelle (*Pipistrellus hesperus*) from which it differs as described in the previous species account. Average external measurements are: total length, 80 mm; tail, 38 mm; hind foot, 7 mm; ear, 13 mm.

Distribution. The western small-footed myotis is known in Texas primarily from the mountainous regions of the Trans-Pecos. It occurs principally in wooded areas, although a few individuals have been taken in grassland and desert scrub habitats. Single specimens from Palo Duro Canyon, Armstrong County, (Hollander

Myotis ciliolabrum. Western Small-footed Myotis. (Merlin D. Tuttle, Bat Conservation International)

and Jones, 1987), and Canyon, Randall County, (Texas Dept. of Health Record; see table 7), document its presence in the High Plains and Rolling Plains of the Texas Panhandle.

Apparently, this bat does not winter in Texas as specimens have only been taken from March through July. Also, the sparse collecting records suggest it may be fairly rare in our state. Outside Texas, the species is found throughout western North America from Canada into Mexico.

Subspecies. Texas specimens are referrable to the subspecies *M. c. ciliolabrum* (Merriam, 1886) as indicated by the most recent taxonomic revision of the species (van Zyll de Jong, 1984).

Life History. Over much of its range in North America, the western small-footed myotis is believed to hibernate within its summer range, but this has yet to be substantiated in Texas. Winter hibernacula are located in rock crevices, caves, and mine tunnels, and this is reportedly one of the last bats to enter torpor. This species has been found roosting in summer beneath slabs of rock, behind loose tree bark, and in buildings. Maternity colonies are small and often located in abandoned houses, barns, or similar structures. One young is born annually from late May to early July.

Western small-footed myotises appear to have feeding and foraging habits similar to those of the California myotis, but there is some suggestion that one species might be a "beetle strategist" and the other a "moth strategist" (Black, 1974). Davis (1974) reports that these bats can take off from the water's surface if knocked

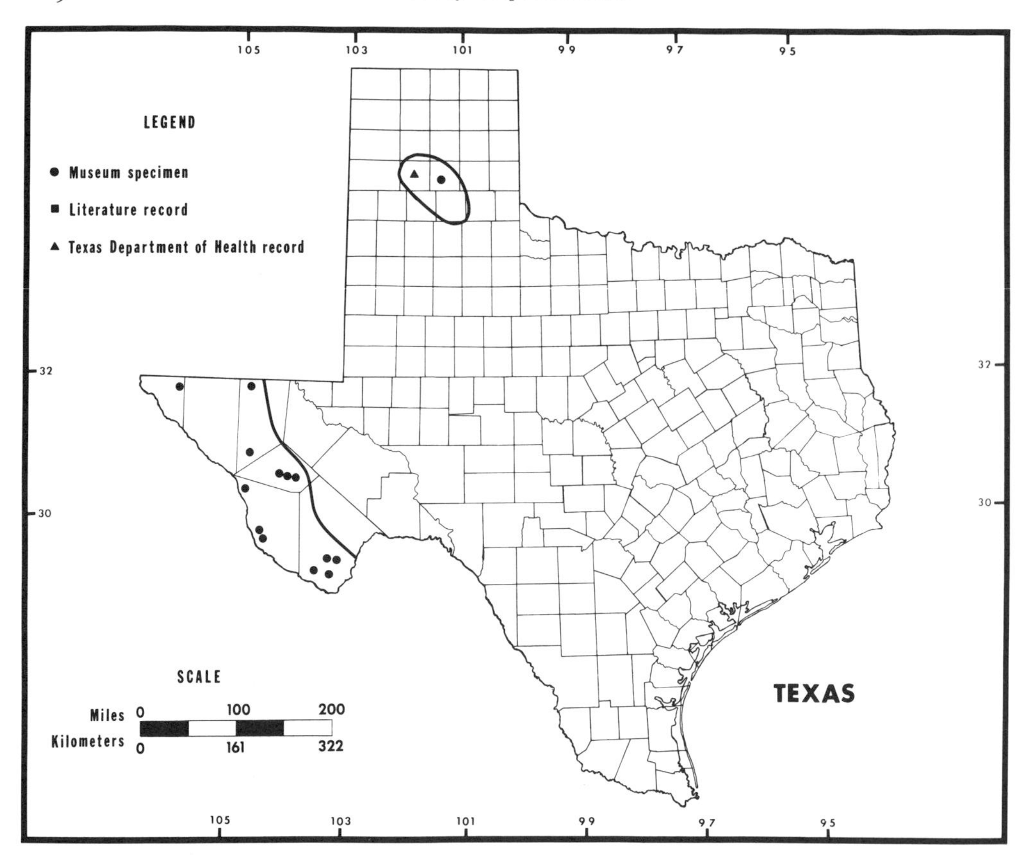

Map 9. Distribution of the western small-footed myotis, *Myotis ciliolabrum ciliolabrum.*

into a pond or tank. A single specimen reported to the Texas Department of Health was not infected with rabies.

Remarks. M. *ciliolabrum* was formerly regarded as a subspecies of *Myotis leibii,* the eastern small-footed myotis, until van Zyll de Jong (1984) elevated it to specific status. M. *ciliolabrum* differs from M. *leibii* in divergent cranial characters and in having a paler pelage, varying from flaxen to yellowish brown, as compared with the pronounced sheen characteristic of M. *leibii.*

Specimens examined (38). *Armstrong Co.:* 29 mi SSW Claude, 1 (TTU). *El Paso Co.:* Hueco Tanks State Park, 1 (UTEP). *Culberson Co.:* McKittrick Canyon, GMNP, 5 (TCWC); Guadalupe Mts., head of Delaware Creek, GMNP, 1 (USNM/FWS); Manzanita Spring, GMNP, 1 (TTU); 16 mi SE Van Horn, 5 (1 USNM/FWS, 4 TCWC); Guadalupe Canyon, 1 (USNM/FWS). *Jeff Davis Co.:* 8 mi S jct. hwys. 166 and 118, 2 (TTU); 3.5 mi NE Ft. Davis, 4 (TTU); DMSP, 2 (TTU); Ft. Davis, old fort ruin, 2 (LACM). *Presidio Co.:* ZH Canyon, Sierra Vieja, 5 (TTU); Chinati Mts., 1 (TTU); Shafter Mine area, Livingston Ranch, 19 mi N Presidio, 1 (ASVRC). *Brewster Co.:* N side Rosillos Mts., 3 (UMMZ); 38 mi S, 14 mi E Marathon, 1 (USNM/FWS); Terlingua Creek, 1 (USNM/FWS); Pine Canyon, BBNP, 1 (BBNHA).

Lasionycteris noctivagans. Silver-haired Bat. (Merlin D. Tuttle, Bat Conservation International)

References. 2, 4, 7, 10, 12, 15, 21, 24, 29, 31, 84, 92, 100, 103, 142, 159, 163, 167, 181, 188, 191, 192, 201, 206, 246, 262, 265, 267, 279, 306, 399, 405.

Lasionycteris noctivagans Le Conte, 1831
Silver-haired Bat

Description. This medium-sized (forearm=37–44 mm) bat is easily recognized by its distinctive pelage coloration. It is entirely black, but individual hairs of the dorsal surface are partially white near the ends, giving the pelage a "frosted" appearance. The ears are short and rounded and the upper surface of the interfemoral membrane is furred, though not as heavily as in the tree bats of the genus *Lasiurus.* Other bats having similar "frosted" pelage include the hoary bat (*Lasiurus cinereus*), which is much larger and grayer in coloration, and the eastern (*L. borealis*) and western (*L. blossevillii*) red bats, which are reddish rather than black. Average external measurements of *Lasionycteris noctivagans* are: total length, 95 mm; tail, 41 mm; hind foot, 9 mm; ear, 14 mm.

Distribution. The silver-haired bat is broadly but erratically distributed across northern North America. It has been recorded from localities scattered throughout Texas and apparently is a fall-spring migrant in our state. Six physiographic regions feature the silver-haired bat—the Pineywoods, Gulf Prairies and Marshes, Edwards Plateau, Rolling Plains, High Plains, and Trans-Pecos. Although primarily a species of forested areas, this bat occasionally may be found in xeric habitats during migration.

Midsummer records of silver-haired bats do not exist for Texas, although the

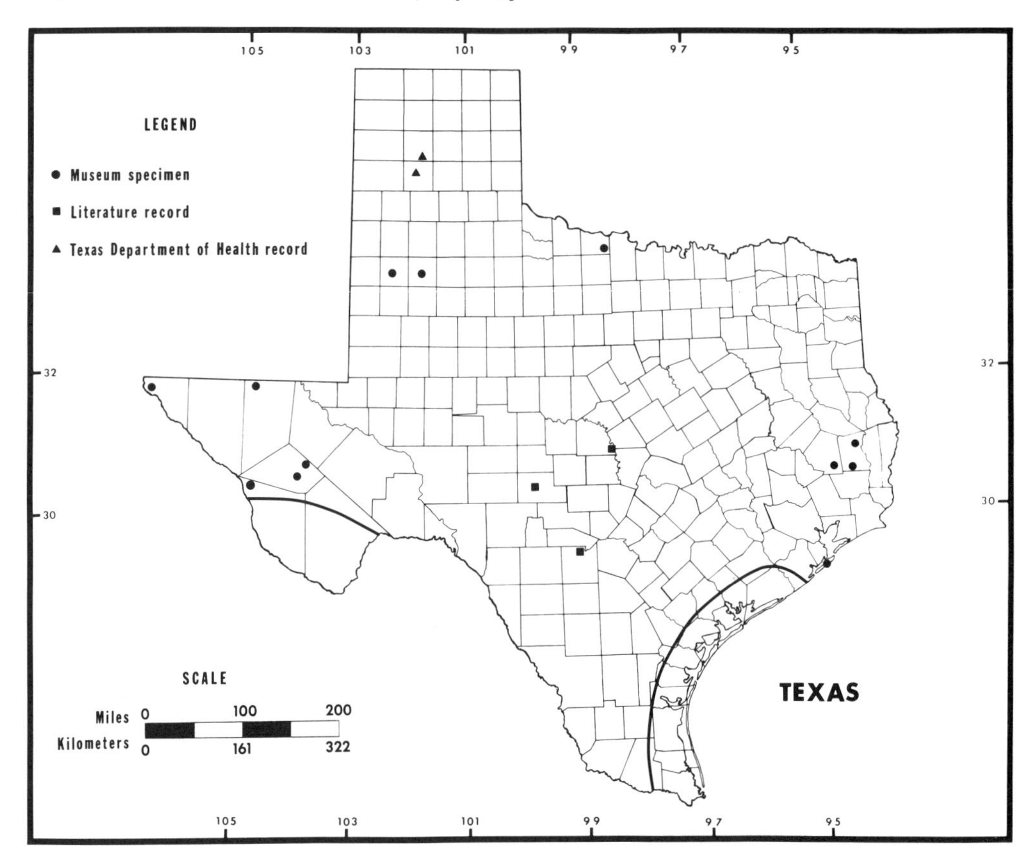

Map 10. Distribution of the silver-haired bat, *Lasionycteris noctivagans.*

Guadalupe Mountains of the Trans-Pecos region may support a summer population. Interestingly, most summer records of this bat across the southwest are of males, suggesting that geographical segregation of the sexes may occur during the warmer months. Females appear to move north in spring and summer to bear young, whereas the males remain behind at more southerly locales.

Subspecies. *L. noctivagans* Le Conte, 1831, is monotypic and no subspecies are recognized.

Life History. *L. noctivagans* is generally considered a "tree bat" which roosts in hollow trees, woodpecker holes, and behind loose bark, but it also has been found in buildings. Although they may occasionally hibernate as far north as New York and British Columbia (Kunz, 1982), most silver-haired bats are believed to overwinter in the southern part of the species' range.

Adult females raise one, or possibly two, young annually on the summer range before they migrate south in fall. Small maternity colonies may form in hollow trees and abandoned bird nests. The young are able to fly at about three weeks of age.

This bat typically forages in or near coniferous and/or mixed deciduous forests, adjacent to ponds or other bodies of water. It is a relatively late flyer, often appearing after other species have begun feeding. As in most other insectivorous bats, *L. noctivagans* is opportunistic in its food habits, taking a variety of small to medium-sized insects, including moths, bugs, beetles, flies, and caddisflies (Kunz, 1982). Captive specimens have eaten banana, bits of raw meat, and insects (Barbour and Davis, 1969).

Five specimens of *L. noctivagans* reported to the Texas Department of Health were not infected with rabies.

Specimens examined (35). *Wichita Co.*: Wichita Falls, 2 (MWSU). *Hockley Co.*: Levelland, 1 (TTU). *Lubbock Co.*: Lubbock, 1 (TTU). *El Paso Co.*: El Paso, 2 (UTEP). *Culberson Co.*: McKittrick Canyon, GMNP, 16 (TCWC); Thrush Hollow, 0.25 mi S Pratt Lodge, S McKittrick Canyon, GMNP, 1 (TTU). *Polk Co.*: 2 mi E, 1.7 mi S Camp Ruby, 1 (TCWC). *Tyler Co.*: 3.6 mi S, 2.9 mi W Town Bluff, 2 (TCWC); 4.2 mi S, 0.8 mi W Warren, Hickory Creek Unit, BTNP, 1 (TCWC). *Jeff Davis Co.*: 10 mi S Balmorhea, 1 (TTU); Davis Mts. Resort, Old McGuire Homestead, 1 (SRSU). *Presidio Co.*: ZH Canyon, 14.5 km SW Valentine, 5 (CCSU). *Galveston Co.*: Galveston Island, west-central residential area, 1 (TCWC).

Additional records: *San Saba Co.*: Gorman Creek, 4 mi SSE Bend (Wilkins et al., 1979). *Kimble Co.*: Texas Tech Univ. Center at Junction (Manning et al., 1987). *Medina Co.*: near Bandera (Davis, 1974); 18 mi W Bandera (Blair, 1952b).

References. 4, 6, 7, 10, 12, 13, 15, 20, 21, 22, 24, 25, 26, 31, 49, 84, 89, 97, 112, 142, 147, 167, 188, 198, 201, 245, 248, 252, 256, 257, 262, 265, 267, 271, 279, 288, 291, 304, 357, 362, 363, 418, 423.

Pipistrellus hesperus (H. Allen, 1864)
Western Pipistrelle

Description. This is the smallest North American bat (forearm = 27–33 mm, weight = 3–6 g). Pelage coloration varies from light gray to yellowish and a dark and leathery facial mask is present across the rostrum and eyes from ear to ear. The wing and tail membranes are also very dark. *P. hesperus* has a keeled calcar, short and rounded ears, and the tragus is slightly curved and blunt at the tip. As with many vespertilionids, males are typically smaller than females of the same age.

P. hesperus may be confused with the California myotis (*M. californicus*) and the western small-footed myotis (*M. ciliolabrum*), both of which are larger and have the narrow, straight, and sharply pointed tragus characteristic of the genus *Myotis*. The eastern pipistrelle (*P. subflavus*) is also similar in appearance to *P. hesperus* and the ranges of these two bats overlap in Texas. Characters distinguishing the two species are given in the account for *P. subflavus*. Average external measurements of *P. hesperus* are: total length, 74 mm; tail, 29 mm; hind foot, 6 mm; ear, 11 mm.

Distribution. The western pipistrelle is one of the most common bats of the desert Southwest. It is particularly abundant in the mountain ranges and rocky can-

Pipistrellus hesperus. Western Pipistrelle. (John L. Tveten)

yon country of Trans-Pecos Texas, but it also is known from scattered localities in the High Plains, Rolling Plains, and the northern and western edges of the Edwards Plateau.

The western pipistrelle is sporadically active during winter in Texas. A few winter records have been documented, and there is no evidence that the species migrates. Apparently these bats do not enter a deep torpor and are capable of arousing and becoming active during warm spells in winter.

Western pipistrelles have been captured at elevations ranging from near sea level on the lower Rio Grande to 2,073 m in the Chisos Mountains of Big Bend National Park. In winter, females are thought to hibernate at higher elevations than males, possibly because their slightly larger size makes them less vulnerable to climatic extremes.

Subspecies. Texas specimens are referrable to the subspecies *P. h. maximus* Hatfield, 1936, as indicated by the latest taxonomic revision of the species (Findley and Traut, 1970).

Life History. Day roosts include rock crevices, under rocks, burrows, and buildings. Similar sites are also used as night roosts and in winter these bats hibernate in mine tunnels and caves. Western pipistrelles tend to roost singly or in very small groups. A maternity colony of twelve individuals is the largest known group of these bats.

These are among the most diurnal of bats, beginning their foraging flights very early in the evening and often remaining active through the early morning hours.

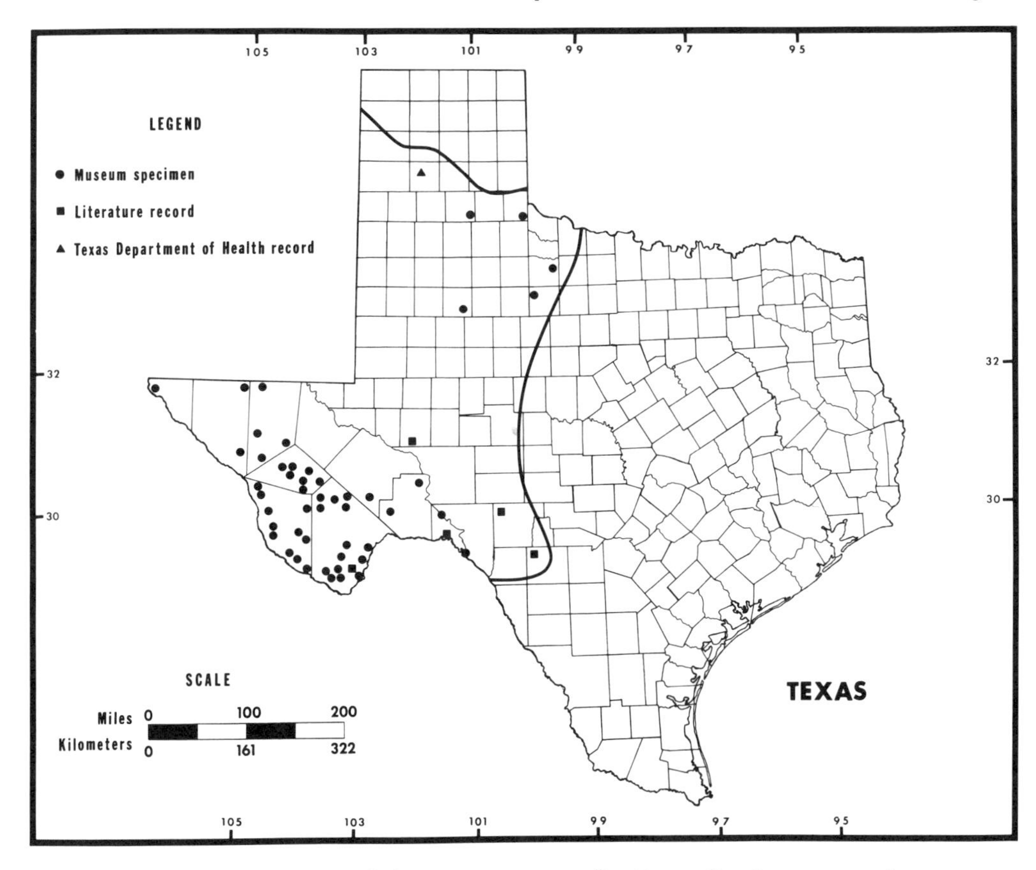

Map 11. Distribution of the western pipistrelle, *Pipistrellus hesperus maximus.*

Pipistrelles are slow bats and may be distinguished on the wing by their slow, fluttery flight, which is restricted to small foraging circuits. Occasionally, individual bats have been observed on the wing during midday, during which time they water to alleviate stress caused by the arid environments they inhabit.

Western pipistrelles forage from 2 to 15 m above ground on small, swarming insects and consume about 20 percent of their body weight in insects per feeding (Hayward and Cross, 1979). Specific prey items include caddisflies, stoneflies, moths, small beetles, leaf and stilt bugs, leafhoppers, flies, mosquitoes, ants, and wasps. In Arizona, flying ants are their principal food item during the summer rainy months. According to Hayward and Cross (1979), stomachs of individual bats often contain only a single species of insect, or if more than one species is present, the remains are clumped together within the stomach, suggesting that these bats take advantage of swarming insects and feed intensely within such swarms.

Parturition occurs from June to July, after a gestation period of approximately forty days. Maternity colonies may be established in buildings or rock crevices, and the females give birth to two young annually. Newborn bats weigh slightly less than one gram at birth but grow quickly, with juveniles becoming volant at about one month of age.

Only one western pipistrelle has been reported to the Texas Department of Health, and that individual was not infected with rabies.

Specimens examined (592). *Briscoe Co.*: 6.1 mi N, 0.1 mi W Quitaque, 2 (TTU). *Childress Co.*: Panther Cave, 22 mi SE Childress, 2 (TTU). *Knox Co.*: 5 mi N Vera, 1 (MSU). *Garza Co.*: Justiceburg, 3 (TTU); 7 mi E Justiceburg, 4 (CCSU). *Haskell Co.*: 10 mi W Rochester, 1 (MWSU). *El Paso Co.*: El Paso, 4 (USNM/FWS); Lomas del Rey, El Paso, 1 (UTEP). *Hudspeth Co.*: 0.56 mi S, 4.3 mi W Guadalupe Peak, GMNP, 1 (TTU); crossroads, GMNP, 5 (TTU); Middle Tank, Wind Canyon, Eagle Mts., 7 (UTEP). *Culberson Co.*: McKittrick Canyon, GMNP, 5,400 ft., 18 (TCWC); 0.1 mi S, 4.3 mi W Guadalupe Peak, GMNP, 2 (TTU); 7 mi N Pine Springs, GMNP, 5,300 ft., 2 (TCWC); Signal Peak, GMNP, 2 (TTU); Smith Springs, GMNP, 1 (TTU); Manzanita Springs, GMNP, 5,500 ft., 1 (TTU); 25 mi N Van Horn, 3,500 ft., 12 (TCWC); 1 mi N Kent, 4,000 ft., 3 (TCWC); 16 mi SE Van Horn, 4,100 ft., 2 (TCWC). *Pecos Co.*: 13 mi N, 19.7 mi E Marathon, 4,300 ft., 5 (CM). *Jeff Davis Co.*: 8 mi W jct. hwys. 166 and 118, 1 (SRSU); 20 mi SW Toyahvale, Davis Mts., 11 (USNM/FWS); 1.5 mi W Mt. Locke, 1 (UMMZ); Limpia Creek, 16 mi NE Ft. Davis, 2 (TCWC); 14 mi NE Ft. Davis, old Whittenburg ranch, 2 (LACM); 5 mi E Mt. Livermore, 1 (UMMZ); 8 mi S jct. hwys. 166 and 118, 5 (TTU); Limpia Canyon, 5 mi N Ft. Davis, 4,800–5,000 ft., 21 (TCWC); 2 mi NW Ft. Davis, 2 (UMMZ); Limpia Canyon, 1 mi N Ft. Davis, 4,800 ft., 7 (4 MVZ, 3 TCWC); 1 mi N Ft. Davis, 15 (1 TTU, 14 UMMZ); Ft. Davis, 1 (UMMZ); Muzquiz Canyon, 17.3 mi N Alpine, 1 (DMNHT); 15 mi N Alpine, 1 (SRSU); 7 mi S Ft. Davis, 2 (DMNHT). *Terrell Co.*: 13 mi S Sheffield, 3 (TNHC); 15 mi S Sheffield, 6 (TNHC); 16 mi S Sheffield, 3 (TNHC); 17 mi S Sheffield, 1 (TNHC); 18 mi S Sheffield, 2 (TNHC); 19 mi S Sheffield, 1 (TNHC); 20 mi S Sheffield, 10 (TNHC); 20 mi SW Sheffield, Independence Creek, 1 (ASVRC); 21 mi S Sheffield, 6 (TNHC); Sanderson, 1 (USNM/FWS); 2 mi E Sanderson, 2,775 ft., 1 (TCWC). *Presidio Co.*: 12 mi W Valentine, 3 (TNHC); 11 mi W Valentine, 23 (TNHC); 9 mi SW Valentine, 2 (CCSU); 10 mi WSW Valentine, 1 (TNHC); 11 mi SW Valentine, 1 (TNHC); ZH Canyon, Sierra Vieja, 2 (TTU); Paisano, 5 (USNM/FWS); 2 mi S Paisano, 5,000 ft., 11 (TCWC); 8 mi NE Candelaria, 5 (TCWC); 12 mi E Ruidosa, Chinati Mts., 5,000 ft., 3 (TCWC); Upper Wild Horse Canyon, Chinati Mts., 3 (SRSU); Pinto Canyon, 45 mi SW Marfa, 14 mi E Ruidosa, 10 (9 TCWC, 1 TTU); Chinati Mts., 3 (1 TTU, 2 USNM/FWS); 37 mi S Marfa, Harper Ranch, 4,000 ft., 6 (TCWC); Bandera Ranch, 1 (MWSU); Bandera Mesa, 13 (MWSU); 1 mi E Bandera Mesa, 1 (MWSU); 5 mi SE Bandera Mesa, 10 (MWSU); 13 mi E Presidio, 1 (MWSU); La Mota Rancho, 63 mi S Marfa, 4 (TNHC); between Lajitas and Redford on Rio Grande, 2 (SRSU); 30 mi SSE Redford, 1 (MWSU); mouth Cottonwood Canyon, 1 (TNHC); no specific locality, 1 (AMNH). *Brewster Co.*: 17.9 mi N, 0.3 mi E Marathon, 2 (TTU); 17.9 mi N, 3 mi E Marathon, 2 (TTU); 13.2 mi N, 2.6 mi E Marathon, 5,200 ft., 1 (TTU); Spicewood Ranch, 17 mi W, 13 mi N Marathon, 2 (UIMNH); Alpine, 2 (USNM/FWS); Woodward Ranch Campground, 12 mi S Alpine, 1 (SRSU); Marathon, 5 (USNM/FWS); 15 mi N BBNP, 3 (TTU); 4.5 mi NE jct. Maravillas Creek and Rio Grande, 3 (DMNHT); BGWMA, 145 (2 TTU, 7 TCWC, 20 TNHC, 101 DMNHT, 15 LACM); 40 mi S Marathon, Tornillo Creek,

1 (AMNH); Ernesta, 50 mi S Marathon, 3 (USNM/FWS); 27 mi N Government Springs, 2,500 ft., 1 (AMNH); Rosillos Mts., 2 (UMMZ); Neville Spring, BBNP, 1 (FMNH); Burnham Ranch, Government Springs, 3,950 ft., BBNP, 8 (AMNH); Terlingua, 5 (USNM/FWS); Grand Canyon, Terlingua, 2 (USNM/FWS); Oak Springs, BBNP, 4,000 ft., 2 (TCWC); Oak Creek, BBNP, 4 (BBNHA); Terlingua Ghost Town, 4 (MWSU); 2 mi E Terlingua, 1 (TTU); 6 mi E Terlingua, 1 (SRSU); 3 mi S Government Springs, 4,500 ft., BBNP, 5 (AMNH); flats NE Chisos Mts., 3,800 ft., BBNP, 3 (FMNH); W base Chisos Mts., 3,500 ft., BBNP, 6 (FMNH); Chisos Mts., The Basin, BBNP, 2 (FMNH); Chisos Mts., BBNP, 5 (4 AMNH, 1 USNM/FWS); Mt. Emory, Chisos Mts., BBNP, 1 (FMNH); Chisos Mts., Juniper Canyon, BBNP, 4 (FMNH); Kibee Spring, Chisos Mts., 5,700 ft., BBNP, 24 (FMNH); Wilson's Tank, 5 mi SE The Basin, BBNP, 1 (FMNH); 2 mi W Boquillas, 2,080 ft., BBNP, 1 (AMNH); Boquillas, BBNP, 10 (1 BBNHA, 2 MVZ, 7 USNM/FWS); 1 mi SW Boquillas, 1,800 ft., BBNP, 2 (AMNH); Glenn Spring, 2,606 ft., BBNP, 1 (MVZ); 81 mi SSE Alpine, 5,300 ft., 1 (MWSU); Nail's Ranch, E side Burro Mesa, BBNP, 3,500 ft., 8 (7 LACM, 1 TCWC); E base Burro Mesa, 3,500 ft., BBNP, 17 (MVZ); Santa Elena Campground, BBNP, 1 (TCWC). *Val Verde Co.:* Pecos Crossing, 4 mi SW Pandale, 1 (MWSU); mouth of Pecos River, 1 (USNM/FWS).

Additional records: *Upton Co.:* 3 mi S, 5 mi E McCamey (Manning et al., 1987). *Brewster Co.:* Paisano (Bailey, 1905); Giant Dagger Yucca Flats, BBNP (Easterla, 1968). *Terrell Co.:* no specific locality (Hermann, 1950). *Val Verde Co.:* no specific locality (Davis, 1974); 0.5 mi E Langtry (Manning et al., 1987). *Edwards Co.:* no specific locality (Davis, 1974). *Uvalde Co.:* no specific locality (Davis, 1974).

References. 1, 2, 3, 4, 6, 7, 10, 12, 15, 21, 24, 29, 33, 65, 69, 76, 78, 84, 92, 93, 99, 103, 105, 155, 163, 167, 181, 188, 196, 198, 201, 207, 213, 221, 224, 226, 245, 256, 257, 265, 267, 279, 288, 304, 306, 342, 356, 358, 362, 393, 414.

Pipistrellus subflavus (F. Cuvier, 1832)
Eastern Pipistrelle

Description. This is a small (forearm=31–35 mm, weight=4–6 g), pale yellowish bat that is characterized by the unique "tricolored" condition of the pelage in which the base of each individual hair is dark, the middle band is lighter, and the tips are dark. This feature readily distinguishes *P. subflavus* from other similar-sized species. The calcar is not keeled, and the leading edge of the wing membranes is paler than the rest of the membrane.

The eastern pipistrelle can be confused with its western congener (*P. hesperus*), from which it is distinguished by its unique, tricolored fur; unkeeled calcar; and paler wing membranes. Also, the tragus in *P. subflavus* is not curved as in *P. hesperus.* Average external measurements are: total length, 77 mm; tail, 35 mm; hind foot, 9 mm; ear, 12 mm.

Distribution. The eastern pipistrelle is known from seven physiographic regions of Texas—the Rolling Plains, Blackland Prairies, Edwards Plateau, South Texas Plains, Post Oak Savannah, Gulf Prairies and Marshes, and Pineywoods—but is

Pipistrellus subflavus. Eastern Pipistrelle. (John L. Tveten)

most common in the eastern portion of the state where it has commonly been collected along bottomland streams and forest flyways. This species also is abundant in caves along the southern and eastern edge of the Balcones Escarpment. It is known to hibernate in caves within its summer range and is a year-round resident of Texas.

Subspecies. Two subspecies occur in Texas, according to the most recent review of the taxonomy of the species (Davis, 1959). *P. s. subflavus* (F. Cuvier, 1832) is the most wide-ranging, occurring in all seven physiographic regions mentioned above. *P. s. clarus* Baker, 1954, which is slightly larger and paler than *P. s. subflavus*, occurs in Val Verde County and adjacent localities in Coahuila, Mexico. These two subspecies are ecologically separable and the geographic area separating the two, including much of the Rio Grande Valley, appears to be uninhabited by the species.

Life History. These bats generally roost singly or in small groups in caves, mines, rock crevices, tree foliage, and occasionally in buildings. Caves, mine tunnels, and rock crevices serve as hibernacula with the bats seemingly preferring the warmer locations within such sites. Individuals may show a high degree of fidelity to a particular roosting site, often returning to the exact same spot after a period of disturbance.

Little is known of their food habits in Texas. In Indiana, eastern pipistrelles have been reported to eat small leafhoppers, ground beetles, flies, moths, and ants (Mumford and Whitaker, 1982). They typically forage at treetop level, often over water, and are not usually found in deep forests or open fields.

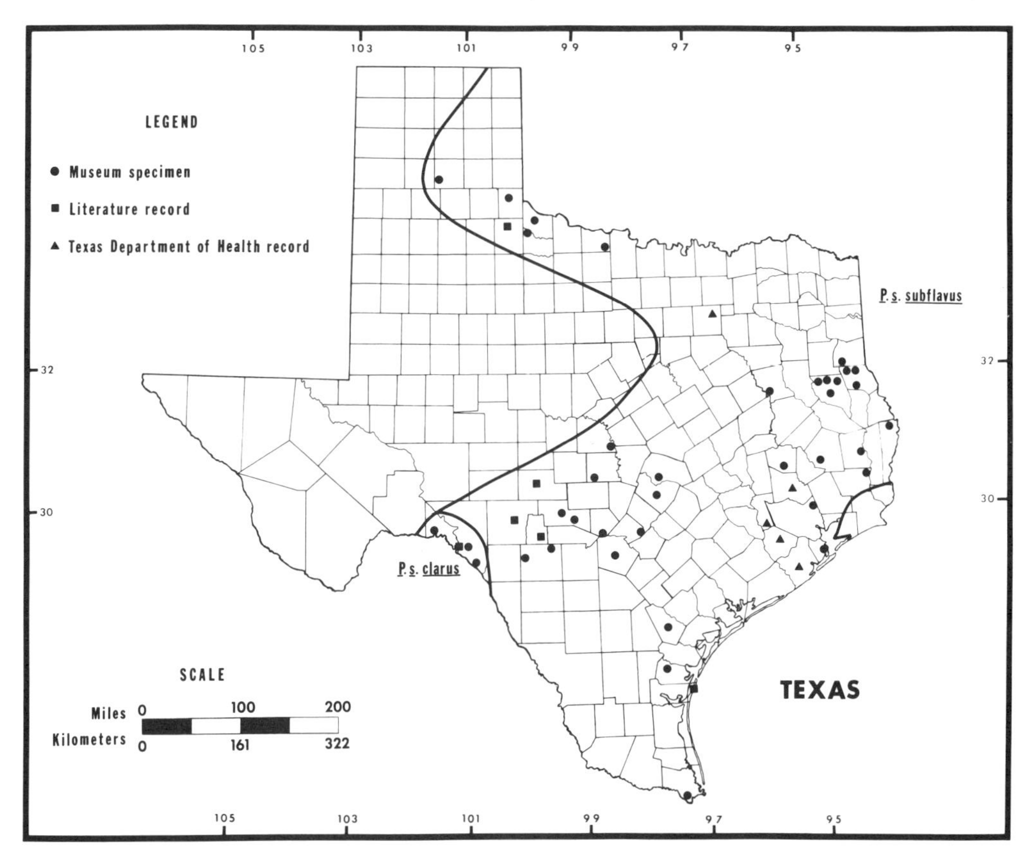

Map 12. Distribution of the two subspecies of the eastern pipistrelle, *Pipistrellus subflavus.*

The young may be born any time from late May to early July. Litter size is usually two, and young bats are able to fly within a month of birth. Maternity colonies are small and often located in open sites that would not be tolerated by most other bats. Mortality is apparently greatest during their first year of life, probably due to heavy losses during winter, and the highest rate of survival is for animals approximately 3.5 years of age (Davis, 1966). Males typically survive longer than females.

Of ten eastern pipistrelles reported to the Texas Department of Health, only one tested positive for the rabies virus.

Specimens examined. Pipistrellus subflavus clarus (6). *Val Verde Co.:* 2 mi W Langtry, Fisher's Fissure, 3 (TTU); Devil's River, 1 (USNM/FWS); Comstock, 1 (USNM/FWS); Del Rio, 1 (USNM/FWS).

Additional records: *Val Verde Co.:* Brotherton-Calk Ranch, 12 mi W Comstock (Reddell, 1968); Comstock (Bailey, 1905).

Pipistrellus subflavus subflavus (85). *Armstrong Co.:* Palo Duro Canyon, 1 (TTU). *Childress Co.:* 25 mi N Childress, 1 (TTU); 16 mi N Childress, 1 (TTU). *Hardeman Co.:* 9 mi W Quanah, 1 (MWSU); 11 mi WSW Quanah, 1 (MWSU); Beezley's Cave, 13 mi SW Quanah, 1 (MWSU); *Foard Co.:* 20 mi SW Quanah, 1 (MWSU).

Wichita Co.: Wichita Falls, 1 (MWSU). *Tarrant Co.*: Arlington, Univ. of Texas at Arlington campus, 1 (UTACV). *Dallas Co.*: Irving, 1 (UTACV). *Panola Co.*: Lake Murvaul, 1 (SFASU); 4 mi SW Gary, 2 (SFASU). *Anderson Co.*: Long Lake, 2 (USNM/FWS); Engeling WMA, Deer Creek Station, 1 (TCWC). *Shelby Co.*: cave between Timpson and Gary, 1 (SFASU); 3 mi N Timpson on Rose Hill Rd., Cave Springs, 1 (SFASU); 15 mi N Center, 1 (TTU); 14 mi N Center, 2 (SFASU); Neuville Cave, 8 mi S Center, 2 (UIMNH); Choice, 6 (LSUMZ). *Nacogdoches Co.*: Hwy 204, 10 mi E Cushing, 1 (SFASU); 1 mi S Cushing, 1 (SFASU); cave, 4 mi SW Garrison, 5 (SFASU); Nacogdoches, 6 (SFASU); 15 mi E Nacogdoches, Garrison hwy., 1 (SFASU). *San Saba Co.*: 5 mi SSE Bend, Gorman Falls Fish Camp, 1 (TCWC). *Newton Co.*: 12 mi N Burkeville, 2 (TCWC); 11.5 mi N Burkeville, 1 (TCWC); 9.3 mi N Burkeville, 1 (TCWC); 8.5 mi N Burkeville, 1 (TCWC). *Polk Co.*: 4 mi E Livingston, 2 (TCWC); 3 mi S Livingston, 1 (TCWC). *Tyler Co.*: 2 mi S, 1.5 mi W Town Bluff, BTNP, 3 (TCWC); 3.6 mi S, 2.9 mi W Town Bluff, 1 (TCWC). *Walker Co.*: 11 mi NW New Waverly, 1 (TNHC). *Llano Co.*: 20 mi N Fredericksburg, 1 (MVZ). *Williamson Co.*: 6 mi W Round Rock, 1 (TNHC). *Travis Co.*: 3 mi W Austin, Bee Cave Rd., 1 (TNHC). *Hardin Co.*: 11 mi N, 2.3 mi E Silsbee, 1 (TCWC). *Kerr Co.*: 5 mi W Hunt, 1 (TCWC); 8 mi SW Kerrville, 3 (TCWC). *Harris Co.*: 4 mi N Huffman, 1 (TCWC). *Kendall Co.*: 4.6 mi SE Boerne, Cascade Caverns, 2 (TCWC); 4.8 mi SE Boerne, 3 (MVZ). *Comal Co.*: 4 mi E Burgheim, 1 (KU); 10 mi E New Braunfels, 1 (TCWC). *Bexar Co.*: San Antonio, 1 (FMNH); Kelly Air Force Base, 1 (KU); Johnson's Cave, 2 (KU). *Uvalde Co.*: 6 mi NW Utopia, 1 (TNHC); cave, 1 mi S Montell, 2 (TCWC). *Galveston Co.*: Clear Creek, 1 (USNM/FWS). *Bee Co.*: 2 mi N Beeville, 1 (CCSU). *Kleberg Co.*: Mile-and-half Plant, 1 (CCSU). *Cameron Co.*: Brownsville, 1 (USNM).

Additional records: *Cottle Co.*: no specific locality (Davis, 1974). *Kimble Co.*: Texas Tech Univ. Center at Junction (Manning et al., 1987); 5 mi S Texas Tech Univ. Center at Junction (Manning et al., 1987). *Edwards Co.*: Rocksprings (Blair, 1952b); Devil's Sinkhole (Manning et al., 1987). *Real Co.*: Leaky (Manning et al., 1987). *Bandera Co.*: no specific locality (Davis, 1974). *Kleberg Co.*: 2.4 km S of N entrance Padre Island Natl. Seashore (Zehner, 1985).

References. 1, 3, 4, 6, 7, 12, 13, 20, 21, 24, 25, 26, 27, 29, 31, 33, 42, 69, 76, 84, 85, 89, 94, 96, 97, 110, 113, 127, 140, 142, 145, 151, 164, 165, 188, 198, 202, 213, 243, 251, 258, 259, 262, 263, 265, 267, 288, 294, 304, 306, 342, 356, 358, 363, 423, 428.

Eptesicus fuscus (Palisot de Beauvois, 1796)
Big Brown Bat

Description. This is a medium to large bat (forearm = 42–51 mm, weight = 13–20 g) with large, broad wings and a broad nose. Pelage coloration varies from pale brown to chestnut above and slightly paler below. The ears are relatively small, leathery, and black; the tragus is broad and rounded. Finally, the calcar is keeled and, as with many other vespertilionids, males are often slightly smaller than females.

The large size and strong, steady flight of the big brown bat distinguishes it

Eptesicus fuscus. Big Brown Bat. (Merlin D. Tuttle, Bat Conservation International)

on the wing. *E. fuscus* is very vocal and often produces an audible chatter during flight. Average external measurements are: total length, 111 mm; tail, 43 mm; hind foot, 11 mm; ear, 17 mm.

Distribution. The big brown bat is distributed throughout the United States except for the extreme southern tip of Florida. It has a somewhat disjunct distribution in Texas, having been recorded only from the eastern, northwestern, and Trans-Pecos parts of the state (Manning et al., 1989). No specimens are available from the central and southern parts of the state, but there is an unverified literature record from Bexar County (see additional records). Although most commonly encountered in summer, this bat apparently is a year-round resident of Texas.

In Big Bend National Park, male and female big brown bats are known to segregate altitudinally in summer (Easterla, 1973). During this season, females are more common in the lowland areas where conditions are more favorable for raising young, and males occupy the higher elevations of the Chisos Mountains.

Subspecies. Two subspecies of big brown bat are known from Texas and they are separated geographically from one another, according to the current taxonomic arrangement of the species (Manning et al., 1989; Burnett, 1983). *E. f. fuscus* (Palisot de Beauvois, 1796) is known from the Pineywoods of East Texas, where it is apparently restricted to pine-oak and longleaf pine vegetation zones (Schmidly et al., 1977). *E f. pallidus* Young, 1908, which is lighter in pelage coloration than *E. f. fuscus*, is known from the High Plains, Rolling Plains, and Trans-Pecos regions of western and northwestern Texas. The subspecies *pallidus* has been collected from wooded, montane areas of the Chisos, Chinati, Davis, Eagle, and Guadalupe mountains and in the Sierra Vieja range of the Trans-Pecos. Records are available throughout winter from caves and canyons of the High Plains and Rolling Plains.

Life History. In summer, big brown bats roost in hollow trees, rock crevices, tunnels, and cliff swallow nests, but they prefer buildings. They seem relatively intolerant of high temperatures at the roost site. When the temperature rises above 33°–35°C, the bats move to a cooler location or abandon the roost entirely. In winter, these bats hibernate in caves, mine tunnels, rock crevices, storm sewers, and buildings. When hibernating in caves, they generally choose sites near the entrance where the temperature is low and the relative humidity is less than 100 percent. When ambient temperatures drop below freezing, the bats will arouse and move to a different, more favorable location. *E. fuscus* roosts singly or in small groups and is often found in association with the Brazilian free-tailed bat (*Tadarida brasiliensis*), the pallid bat (*Antrozous pallidus*), the Yuma myotis (*Myotis yumanensis*), and the little brown myotis (*Myotis lucifugus*).

Big brown bats feed upon a variety of medium-sized (6–12 mm long) night-flying insects, but curiously they do not appear to prey upon moths to an appreciable degree. Mumford and Whitaker (1982) analyzed the stomachs of 184 *E. fuscus* from Indiana and found that ground beetles formed the greatest volume of

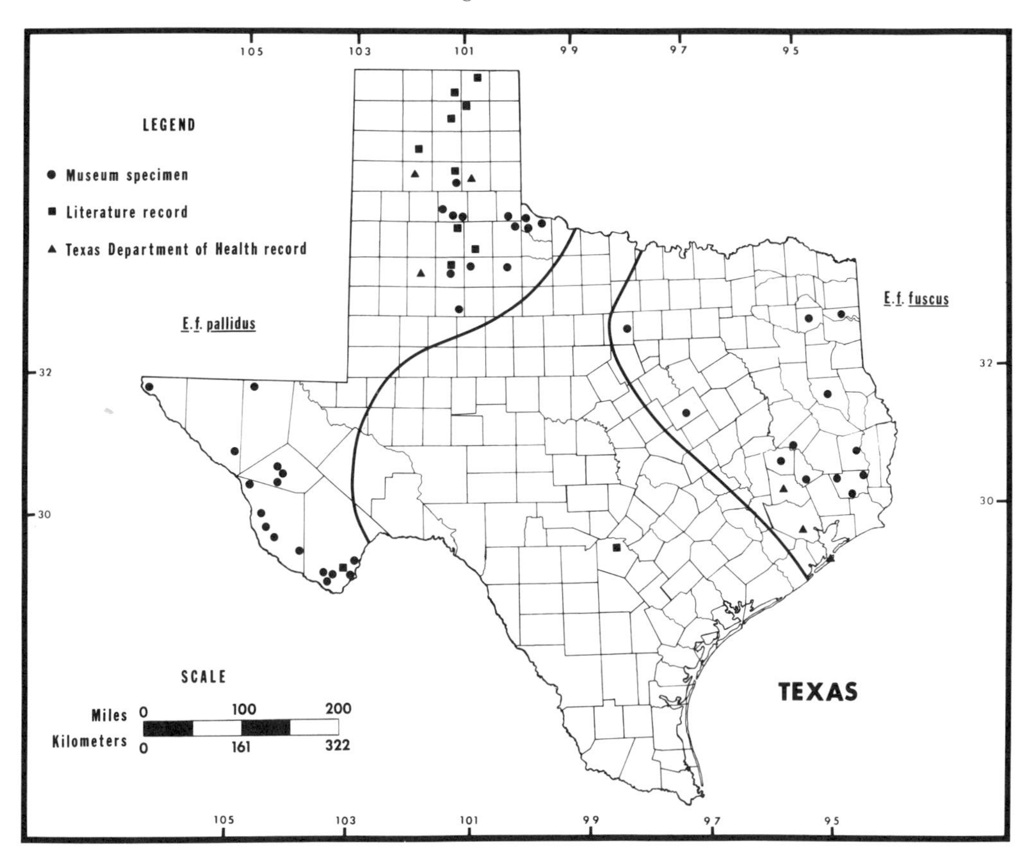

Map 13. Distribution of the two subspecies of the big brown bat, *Eptesicus fuscus.*

food. Scarab beetles were second in abundance, followed by the spotted cucumber beetle. Overall, beetles formed about 50 percent of the entire volume of food in their sample. Ross (1967) analyzed the digestive tract of *E. f. pallidus* from Arizona and Mexico and was able to document the occurrence of termites, true bugs, leafhoppers, June beetles, leaf beetles, and flying ants. Big brown bats seem to prefer foraging among tree foliage rather than below or above the forest canopy. Also, these bats can swim if necessary, but they cannot take off from the water's surface.

The time of parturition varies with latitude, but in Texas most births occur from late May to June. Interestingly, big brown bats in the eastern part of the United States (subspecies *fuscus*) usually produce two young per litter, whereas in the Rocky Mountains and westward (subspecies *pallidus*) the litter size is one. Since Texas spans both of these ranges, it must be assumed that bats in the Trans-Pecos have a litter size of one, whereas those in the Pineywoods typically produce twins. Indeed, six gravid females from the Trans-Pecos each carried a single fetus, in keeping with their assignment to *E. f. pallidus*, but no fetal counts are currently available for bats from eastern Texas or from the northwest (Manning et al., 1989).

At birth, young *E. fuscus* weigh 2.7–3.6 g and grow quickly, gaining as much as 0.5 g per day (Kunz, 1974). Maternity colonies are often located in buildings and may contain from twenty to three hundred individuals. Adult males are usually not present in maternity colonies until the young mature, when males may begin using the maternity colonies more frequently. At four weeks of age the young bats begin foraging for themselves and reach adult size approximately two months after birth.

Predators include barn owls, horned owls, rat snakes, and kestrels (Barbour and Davis, 1969; Davis, 1974). Avoiding these hazards, big brown bats are long-lived with a record of eighteen years known for one banded individual (Hitchcock, 1965). Nineteen big brown bats have been reported to the Texas Department of Health, of which only one proved rabid.

Specimens examined. Eptesicus fuscus fuscus (34). *Palo Pinto Co.*: Brazos, 1 (USNM/FWS). *Upshur Co.*: Gilmer, 13 (TCWC). *Marion Co.*: Jefferson, 1 (USNM/FWS). *Nacogdoches Co.*: Stephen F. Austin State Univ. campus, 1 (TTU). *McLennan Co.*: Waco, 1 (SM). *Trinity Co.*: Trinity, 2 (1 TCWC, 1 USNM/FWS). *Tyler Co.*: 1.1 mi S, 1 mi W Town Bluff, 1 (TCWC); 2 mi S, 1.5 mi W Town Bluff, BTNP, 1 (TCWC); 3.8 mi N, 1.9 mi W Spurger, BTNP, 2 (TCWC); Beech Creek Unit, BTNP, 0.6 mi N, 0.7 mi W Spurger, 2 (TCWC). *Walker Co.*: 2 mi SW Huntsville, 2 (TCWC). *San Jacinto Co.*: jct. farm rd. 945 and E San Jacinto River, 1 (TTU); 5 mi NW Cleveland, 3 (TTU). *Hardin Co.*: Grady, 1 (USNM/FWS); 11 mi N, 2.3 mi E Silsbee, BTNP, 1 (TCWC); 7 mi NE Sour Lake, 1 (USNM/FWS).

Eptesicus fuscus pallidus (179). *Armstrong Co.*: 29 mi SW Claude, 1 (TTU). *Briscoe Co.*: 22 mi E Tulia, 1 (UMMZ); 15 mi E Silverton, 2 (TTU); Los Lingos Canyon, 4 (TTU); 6 mi N, 1 mi W Quitaque, 4 (TTU). 6 mi S Quitaque, 1 (UMMZ). *Childress Co.*: 12 mi S Childress, 1 (MWSU). *Hardeman Co.*: 3 mi SE Lazare, 1 (MWSU); 7 mi E Quanah, 1 (MWSU); 12 mi E Quanah, 1 (MWSU); 11 mi WSW Quanah, 2 (MWSU); 11 mi SW Quanah, 1 (MWSU); Beezley's Cave, 13 mi SW Quanah, 5 (MWSU). *Cottle Co.*: 22 mi SE Childress, Panther Cave, 1 (TTU). *Crosby Co.*: 5 mi E Silver Falls, 1 (TTU). *Dickens Co.*: McAdoo, 1 (TTU). *King Co.*: River Styx Cave, 10 mi NE Guthrie, 1 (TTU). *Garza Co.*: Salt Fork of the Brazos, 3 mi E Justiceburg, 4 (TTU); 7 mi E Justiceburg, 3 (CCSU). *El Paso Co.*: Fort Bliss, 27 (UTEP). *Hudspeth Co.*: Wind Canyon, Eagle Mts., 3 (1 UTEP, 2 SRSU). *Culberson Co.*: The Bowl, GMNP, 4 (TCWC); jct. N McKittrick Canyon and Devil's Den Canyon, GMNP, 1 (TTU); McKittrick Canyon, GMNP, 17 (TCWC); Thrush Hollow, 0.25 mi S Pratt Lodge, S McKittrick Canyon, GMNP, 1 (TTU); Pine Springs, GMNP, 2 (1 TCWC, 1 TTU). *Jeff Davis Co.*: 8 mi W jct. hwys. 166 and 118, 4 (SRSU); 8 mi S jct. hwys. 166 and 118, 1 (TTU); Findley's Ranch, 15 mi E Valentine, 1 (USNM/FWS). *Presidio Co.*: 11 mi W Valentine, 2 (TNHC); Sierra Vieja, 9 mi SW Valentine, 3 (TTU); 8 mi NE Candelaria, 4 (TCWC); Pinto Canyon, 45 mi SW Marfa, 14 mi SE Ruidosa, 13 (1 UIMNH, 11 TCWC, 1 TTU); Chinati Mts., 12 mi E Ruidosa, 5,000 ft., 6 (TCWC); Wild Horse Canyon, Chinati Mts., 14 (SRSU); Livingston Ranch, 19 mi N Presidio, 2 (ASVRC); 7 mi NE Bandera Mesa, 2 (MWSU). *Brewster Co.*: BGWMA, 23 (DMNHT); Chisos Mts., BBNP, 1 (USNM/FWS); Basin Sewage Lagoon, BBNP, 1 (BBNHA); Oak Creek, 4,000 ft., BBNP, 1 (TCWC); Terlingua Creek, 6 mi E Terlingua, 3 mi W

Study Butte, 3 (SRSU); Rio Grande Village, BBNP, 1 (TCWC); Nail's Ranch, E Side Burro Mesa, 3,500 ft., BBNP, 6 (1 MVZ, 5 TCWC).

Additional records: *Hansford Co.:* Spearman (Jones et al., 1988). *Ochiltree Co.:* Perryton (Jones et al., 1988). *Hutchinson Co.:* 9 mi E Stinnett (Jones et al., 1988). *Roberts Co.:* 10 mi S, 15 mi E Spearman (Jones et al., 1988). *Hemphill Co.:* no specific locality (Davis, 1974). *Potter Co.:* 16 mi N Amarillo (Jones et al., 1988); 15 mi N Amarillo (Jones et al., 1988). *Armstrong Co.:* Goodnight (Strecker, 1910). *Hardemann Co.:* Acme (Dalquest, 1968); Walkup Cave, 25 mi E Childress (Milstead and Tinkle, 1959). *Floyd Co.:* 7 mi SW Quitaque (Milstead and Tinkle, 1959). *Motley Co.:* Roaring Springs (Milstead and Tinkle, 1959). *Crosby Co.:* Ralls (Milstead and Tinkle, 1959); McAdoo (George and Strandtmann, 1960); English Ranch (George and Strandtmann, 1960). *Brewster Co.:* Giant Dagger Yucca Flats, BBNP (Easterla, 1968). *Bexar Co.:* Camp Bullis (Brennan, 1945).

References. 1, 2, 3, 4, 6, 7, 10, 12, 13, 15, 20, 21, 22, 24, 25, 26, 27, 29, 31, 33, 76, 84, 89, 91, 92, 93, 98, 99, 102, 103, 105, 106, 108, 110, 114, 142, 145, 147, 151, 154, 155, 156, 158, 163, 167, 181, 188, 198, 201, 202, 216, 243, 245, 255, 256, 257, 262, 265, 267, 268, 277, 279, 289, 293, 294, 304, 306, 322, 342, 356, 358, 362, 363, 374, 388, 389, 396, 400, 401, 402, 414, 423.

Lasiurus borealis (Müller, 1776)
Eastern Red Bat

Description. The eastern red bat is a medium-sized (forearm = 35–45 mm, weight = 10–15 g), distinctly reddish or orange-red bat with short, rounded ears and a relatively long tail. The interfemoral membrane is densely and completely furred dorsally. The tragus is triangular in shape and has a slight forward bend to the tip. Pelage coloration is sexually dimorphic: females have reddish fur with white hair tips that produce a frosted appearance; males lack the white hairs and are more uniformly red. Average external measurements for *L. borealis* are: total length, 106 mm; tail, 45 mm; hind foot, 8 mm; ear, 13 mm.

The eastern red bat is similar in appearance to the Seminole bat (*Lasiurus seminolus*), but coloration will distinguish between the two species. *L. seminolus* has a rich, mahogany brown pelage in contrast to the brick red fur of *L. borealis. L. borealis* and the newly recognized western red bat (*Lasiurus blossevillii*) are also very similar and may be confused in Trans-Pecos Texas where they could occur together. *L. blossevillii* is rusty red to brownish in coloration and lacks the frosted appearance of *L. borealis.* Also, *L. blossevillii* is significantly smaller in most cranial measurements (Schmidly and Hendricks, 1984), and the posterior third of the interfemoral membrane is only sparsely haired.

Distribution. The eastern red bat is a tree-roosting, forest-dwelling species. It is widely distributed throughout Texas and is one of the most common bats of the eastern part of the state (Schmidly et al., 1977). There are very few records from the Trans-Pecos, where the species seems to be limited primarily to mountainous areas.

Lasiurus borealis. Eastern Red Bat. (Merlin D. Tuttle, Bat Conservation International)

L. borealis is highly migratory and, although it is considered a year-round resident of eastern Texas, collecting records decline sharply in the winter months. This species is thought to be only a summer migrant in the Trans-Pecos region. The status of the species in Big Bend National Park is an enigma. Two specimens have been reported from that area but both records are suspect (Easterla, 1975). One specimen of eastern red bat has been taken just north of the park boundary, however (Smith, 1975).

Subspecies. *L. borealis* (Müller, 1776) is a monotypic species with no subspecies recognized.

Life History. Bats of the genus *Lasiurus* are tree-roosting species and are almost never found using caves, mines, or similar sites frequented by other bats. Roosting sites are commonly in tree foliage or clumps of Spanish moss where the bats are concealed because they closely resemble dead leaves. These sites, typically found along forest edges and fencerows and situated 1.5–6.0 m above ground, are selected to provide cover and shade from above and both sides. The area directly below the roost must be unobstructed to allow the bats an easy drop for take-off.

Eastern red bats are solitary in their habits except for small family groups consisting of an adult female and her offspring. Their dense fur and ability to lower bodily functions dramatically during cold weather allow them to survive rigorous winter weather in relatively open roosting sites. In fact, the red bat is so adapted to withstanding inclement weather that it may not be able to arouse spontane-

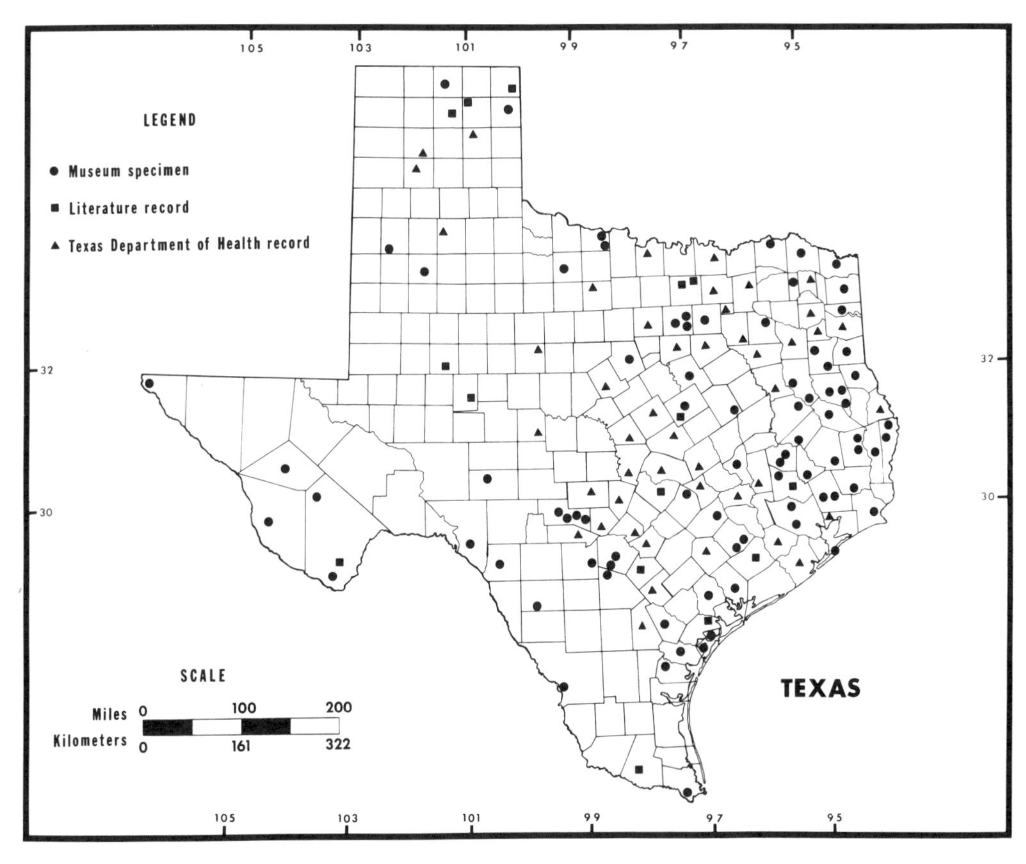

Map 14. Distribution of the eastern red bat, *Lasiurus borealis.*

ously from torpor in caves and thus may not survive hibernation in such sites (Barbour and Davis, 1969).

Eastern red bats typically follow a specific territory while feeding and generally forage near the forest canopy at or above treetop level. They often hunt around street lamps in towns and occasionally land to capture insects. Crepuscular flying insects, such as moths, scarab beetles, planthoppers, flying ants, leafhoppers, ground beetles, and assassin bugs are among their favorite prey items (Mumford and Whitaker, 1982; Ross, 1967).

At northern latitudes breeding takes place in August and September; sperm is stored in the uterus and oviducts through winter, and fertilization and parturition occur in spring. Such "delayed fertilization" is typical of many vespertilionid bats. This same reproductive chronology may also occur in eastern Texas, although copulation probably occurs in the spring since these bats are present and active in that region throughout the year (Schmidly, 1983). The female gives birth to one to four young in May, June, or July (Davis, 1974) after a gestation period of eighty to ninety days. Each baby bat weighs approximately half a gram at birth and is completely hairless. The young are able to fly at three to four weeks and are weaned at five to six weeks of age (Shump and Shump, 1982).

The eastern red bat has been reported to the Texas Department of Health (TDH) more often than any other species of bat. This is no doubt due to the species' wide range, abundance, and propensity to roost and forage around human habitations. During the period from February, 1984, through February, 1987, 626 *L. borealis* were reported to the TDH, of which 46 tested positive for the rabies virus. This represents about a 7 percent infection rate.

Specimens examined (522). *Hansford Co.*: Gruver, 1 (KU). *Hemphill Co.*: Washita River, 4 (ANSP). *Lamb Co.*: 15 mi SW Littlefield, 1 (TTU). *Wichita Co.*: Red River Valley, 9 mi NW Burkburnett, 3 (MWSU); Burkburnett, 2 (MWSU); Wichita Falls, 29 (28 MWSU, 1 USNM/FWS). *Clay Co.*: Henrietta, 1 (MWSU). *Lamar Co.*: Arthur City, 3 (USNM/FWS); Paris, 3 (USNM/FWS). *Red River Co.*: Clarkville, 1 (USNM/FWS). *Lubbock Co.*: Texas Tech Univ. Campus, Lubbock, 1 (TTU); Lubbock, 3 (1 TTU, 2 UMMZ). *Baylor Co.*: Seymour, 1 (MWSU); *Bowie Co.*: 8.0 mi N New Boston, 1 (TCWC). *Franklin Co.*: 6 mi N Mt. Vernon, 8 (LSUMZ). *Cass Co.*: 2 mi E Domino, 2 (TCWC). *Tarrant Co.*: Hurst, 1 (UTACV); Fort Worth, 83 (1 DMNHT, 81 FWMSH, 1 TWC); Arlington, 8 (FWMSH); NTAC Farm, Arlington, 1 (TCWC); Univ. of Texas at Arlington campus, Arlington, 500 ft., 1 (TCWC); no specific locality, 1 (UTACV). *Dallas Co.*: Dallas, 5 (3 ANSP, 1 USNM/FWS, 1 UTACV); no specific locality, 2 (DMNHT). *Marion Co.*: Jefferson, 2 (USNM/FWS). *Van Zandt Co.*: 1 mi SW Grand Saline, 1 (TTU). *Smith Co.*: Tyler, 1 (TCWC). *Erath Co.*: Stephenville, 500 ft., 1 (TCWC). *Rusk Co.* 1.6 mi NE New London, 2 (SFASU); 6 mi E Mt. Enterprise, 1 (SFASU). *Panola Co.*: 7.5 mi ENE Carthage, 4 (TCWC); 6 mi E Carthage, 1 (SFASU); *Hill Co.*: 5.8 mi S, 3.4 mi W Hillsboro, 1 (TCWC). *Cherokee Co.*: Maydelle, 1 (SFASU); 3 mi W Forest, 3 (2 SFASU, 1 TTU). *El Paso Co.*: 8 mi E, 5 mi S El Paso City Hall, 3,700 ft., 1 (KU); Fort Bliss, 1 (USNM/FWS). *Shelby Co.*: Choice 1 (LSUMZ). *Nacogdoches Co.*: Martinsville, 1 (SFASU); Nacogdoches, 2 (SFASU); SFA Campus, 10 (5 SFASU, 5 TTU); SFA Experimental Forest, 2 (TTU); 6 mi E Nacogdoches, 1 (SFASU); Sam Rayburn Reservoir, 1 (SFASU). *McLennan Co.*: Lake Waco, 1 (SM); Waco, 6 (SM); no specific locality, 2 (SM). *Limestone Co.*: 7.8 mi S Personville, 1 (TCWC). *Houston Co.*: 6.3 mi N Ratcliff, 2 (UTACV). *Angelina Co.*: 23 mi S Nacogdoches, 1 (TTU); 5 mi N Lufkin, 1 (SFASU); 1 mi S Lufkin, 1 (SFASU). *Trinity Co.*: Trinity, 2 (1 TCWC, 1 USNM/FWS); 1.3 mi E Trinity, 1 (TCWC). *Newton Co.*: 11.5 mi N Burkeville, 2 (TCWC); 8.5 mi N Burkeville, 4 (TCWC); 7.5 mi N Burkeville, 1 (TCWC); 7.0 mi N Burkeville, 1 (TCWC); Newton, 1 (TTU). *Jasper Co.*: Bouton Lake, 1 (UTACV). *Polk Co.*: 2 mi E, 1.7 mi S Camp Ruby, 1 (TCWC); Big Sandy Unit, BTNP, 2 (TCWC); 4 mi W, 0.3 mi S Dallardsville, 1 (TCWC). *Tyler Co.*: 1.1 mi S, 1.0 mi W Town Bluff, BTNP, 2 (TCWC); 2 mi S, 1.5 mi W Town Bluff, BTNP, 1 (TCWC); 3.6 mi S, 2.9 mi W Town Bluff, 3 (TCWC); 11.7 mi E, 3.6 mi S Woodville, BTNP, 1 (TCWC); 12 mi E, 4 mi S Woodville, BTNP, 1 (TCWC); 3.8 mi N, 1.9 mi W Spurger, BTNP, 7 (TCWC); 0.6 mi N, 0.7 mi W Spurger, 5 (TCWC). *Jeff Davis Co.*: 8 mi S jct. hwys. 166 and 118, Sawtooth Mt., 1 (TTU). *Brazos Co.*: Bryan, 3 (TCWC); vicinity hwy. 21, College Station, 1 (TCWC); 4.0 mi W College Station, 1 (TCWC); College Station, 11 (TCWC); 6.5 mi SW College Station, 300 ft., 1 (TCWC); 6.5 mi SE College Station, 4 (TCWC). *Walker Co.*: 2 mi NE Huntsville, 2 (TCWC); Huntsville, 1 (TCWC); 16 mi SW Huntsville, 1 (TCWC). *San Jacinto Co.*: jct. farm rd. 945 and E San Jacinto River, 32 (TTU); 5 mi NW Cleveland, 10 (TTU). *Sutton Co.*: Sonora, 1 (TTU). *Montgomery Co.*: 20 mi SW Huntsville, 2 (TNHC). *Presidio Co.*: Chinati Mts., 14 mi E Ruidosa, 2 (TTU). *Brewster Co.*: Sul Ross Univ. campus, Alpine, 1 (SRSU); Mt. Emory, Chisos Mts., 2 (FMNH). *Travis Co.*: Austin, 2 (TNHC). *Hardin Co.*: 11.0 mi N, 2.3 mi E Silsbee, 1 (TCWC); 10.9 mi N, 1.8 mi E Silsbee, BTNP, 1 (TCWC); 10.5 mi N, 3 mi E Silsbee, 1 (TCWC); 9.5 mi E, 2 mi N Saratoga, 1 (TCWC); 7 mi NE Sour Lake, 1 (USNM/FWS); Sour Lake, 2 (USNM/ FWS). *Liberty Co.*: 20 mi NW Liberty, 1 (USNM/FWS); 12 mi N Dayton, 3 (TTU); Tarkington Prairie, 1 (USNM/FWS). *Bastrop Co.*: 16 mi NW Giddings, 2 (TTU). *Val Verde Co.*: Devil's River, 1 (USNM/FWS). *Kerr Co.*: Ingram, 4 (1 UMMZ, 3 USNM/FWS); Texas Lions Camp, Kerrville, 1 (TTU); 8 mi SW Kerrville, 1 (TCWC); 8 mi SW Ingram, 2 (TCWC); 13 mi W Hunt, Kerr Wildlife Mgt. Area, 1 (TCWC); 20 mi SW Hunt, 5 (TCWC); 4 mi NE Center Point, 2 (TCWC); 7 mi W Camp Verde, 2 (USNM/FWS). *Fayette Co.*: Warrenton, 1 (TTU). *Jefferson Co.*: Port Arthur, 12 (6 SFASU, 6 TTU). *Harris Co.*: Westfield, 1 (SFASU); Houston, 4 (TCWC). *Colorado Co.*: Eagle Lake, 1 (KU); Garwood, 2 (MSU). *Bexar Co.*: San Antonio, 3 (1 AMNH, 2 FMNH); Somerset, 2 (TNHC). *Kinney Co.*: Fort Clark, 2 (USNM/FWS). *Medina Co.*: LaCoste, 1 (WMM). *Galveston Co.*: Bolivar Peninsula, 7.5 mi NE ferry landing, 1 (TNHC). *Atascosa Co.*: 7 mi SW Somerset, 1 (TNHC). *Jackson Co.*: Lolita, 1 (TTU). *Victoria Co.*: Victoria, 1 (USNM/FWS). *Zavala Co.*: Crystal City, 1 (TAIU). *Bee Co.*: Beeville, 1 (TNHC). *Aransas Co.*: Aransas Refuge, 1 (TCWC); Rockport, 3 (2 AMNH, 1 WMSA). *Webb Co.*: Laredo, 1 (USNM/FWS). *San Patricio Co.*: Welder Wildlife Refuge, 1 (USNM/FWS); Nueces Bay, 1 (USNM/FWS). *Kleberg Co.*: Kingsville, 2 (TAIU). *Cameron Co.*: Brownsville, 111 (7 AMNH, 1 ANSP, 103 USNM/FWS).

Additional records: *Hansford Co.*: Gruver (Jones et al., 1988). *Lipscomb Co.*: Tyson Ranch, 8 mi NW Higgins (Hollander et al., 1987; Jones et al., 1988). *Hutchinson Co.*: 9 mi E Stinnett (Blair, 1954; Jones et al., 1988). *Roberts Co.*: 10 mi S, 15 mi E Spearman (Jones et al., 1988). *Denton Co.*: Pilot Point (Wiseman et al., 1962); Denton (Wiseman et al., 1962). *Dallas Co.*: Dallas (Wiseman et al., 1962). *Howard Co.*: no specific locality (Davis, 1974). *Sterling Co.*: 4 mi S, 4 mi E Sterling City (Manning et al., 1987). *McLennan Co.*: no specific locality (Davis, 1974). *Montgomery Co.*: Conroe (Wiseman et al., 1962). *Brewster Co.*: 10.5 mi N Panther Junction (Smith, 1975); Emory Peak, BBNP (Easterla, 1975). *Travis Co.*: Austin (Wiseman et al., 1962). *Kerr Co.*: 16 mi S Kerrville (Muliak, 1943). *Harris Co.*: Houston (Wiseman et al., 1962). *Real Co.*: Leaky (Manning et al., 1987). *Bexar Co.*: San Antonio (Wiseman et al., 1962). *Wharton Co.*: Wharton (Wiseman et al., 1962). *Wilson Co.*: Floresville (Wiseman et al., 1962). *Refugio Co.*: no specific locality (Davis, 1974). *Nueces Co.*: no specific locality (Davis, 1974). *Hidalgo Co.*: Edinburg (Muliak, 1943). *Cameron Co.*: no specific locality (Davis, 1974).

References. 1, 2, 3, 4, 6, 7, 10, 12, 13, 20, 21, 22, 24, 25, 26, 27, 29, 31, 33, 54, 64, 71, 73, 76, 80, 82, 84, 89, 96, 97, 98, 102, 113, 132, 135, 142, 145, 147, 182, 188, 198,

218, 245, 247, 250, 256, 257, 262, 265, 266, 267, 268, 280, 288, 294, 304, 314, 321, 322, 327, 339, 355, 356, 363, 376, 385, 389, 419, 425.

Lasiurus blossevillii (Lesson and Garnot, 1826)
Western Red Bat

Description. This medium-sized (forearm = 40 mm) bat is similar in overall appearance to the eastern red bat (*Lasiurus borealis*). It has short, rounded ears and a relatively long tail. Pelage coloration is rusty red to brownish and lacks the white-tipped hairs which give the frosted appearance so characteristic of *L. borealis.* The posterior third of the interfemoral membrane is bare or only sparsely haired. *L. blossevillii* is slightly smaller than *L. borealis* and most cranial measurements (i.e., greatest length of skull, zygomatic breadth, mastoid breadth, and length of maxillary toothrow) are significantly smaller. Average external measurements for *L. blossevillii* are: total length, 103 mm; tail, 49 mm; hind foot, 10 mm; ear, 13 mm.

Distribution. *L. blossevillii* ranges across the southwestern and far western areas of the United States south into Mexico and Central America. Only one specimen has been recorded in Texas (Genoways and Baker, 1988) from the Sierra Vieja Mountains in Presidio County of the Trans-Pecos region. Specimens of *L. borealis* have been reported from northwestern Chihuahua (Bogan and Williams, 1970; Anderson, 1972); El Paso County (Jones and Lee, 1962), and the Chisos Mountains, Brewster County, Texas (Anderson, 1972). This gives a broad range of potential overlap between the geographic ranges of the two species in northern Chihuahua and western Texas.

In the southwest, the western red bat is known from scattered localities in New Mexico (Findley et al., 1975) and Arizona (Hoffmeister, 1986). These writers indicate that *L. blossevillii* is occasionally captured in riparian habitats dominated by cottonwoods, oaks, sycamores, and walnuts and is rarely found in desert habitats. This bat has been captured in riparian, xeric thorn scrub, and pine-oak forest habitats of the San Carlos Mountains of Tamaulipas, Mexico, which is approximately 160 km south of the Texas border (Schmidly and Hendricks, 1984). Schmidly and Hendricks (1984) found evidence to suggest that the two species interbreed in northeastern Mexico; however, no data are available to suggest that such intergradation occurs in Texas (Genoways and Baker, 1988).

L. blossevillii is probably migratory in the Southwest. Hoffmeister (1986) reported that specimens from Arizona have been captured only in summer (July and August). Findley et al. (1975) report that specimens in New Mexico have been captured only from May 15 to August 4, and these authors suggest that a winter withdrawal of New Mexico red bats to the southwest seems likely. The Texas specimen was captured on July 15 (Genoways and Baker, 1988).

Subspecies. According to Genoways and Baker (1988) the Texas specimen is referrable to the subspecies *L. b. teliotis* (H. Allen, 1891).

Lasiurus blossevillii. Western Red Bat. (Bruce J. Hayward)

Life History. As stated above, western red bats seem to prefer riparian areas where they roost in tree foliage. Hargrave (1944) reported one specimen found hanging from a fig leaf, about 2.1 m from the ground, at the north edge of Benson, Arizona, near a pond and grove of cottonwood trees. The specimen from Texas was an adult female captured over permanent water in desert scrub habitat in ZH Canyon of the Sierra Vieja.

The food habits and reproductive biology of *L. blossevillii* are poorly documented. Schmidly and Hendricks (1984) reported capturing a female pregnant with three fetuses on May 23. Findley et al. (1975) captured pregnant western red bats in New Mexico on May 15 and June 15, 25, 26, and 29, and lactating females from June 13 to July 12. These observations suggest that *L. blossevillii* may raise as many as three young annually with parturition occurring from mid-May through late June.

Remarks. Based upon genetic studies, Baker et al. (1988) combined and elevated the subspecies *Lasiurus borealis teliotis* and *Lasiurus borealis frantzii* to specific status under the scientific name *Lasiurus blossevillii,* which they distinguished from the eastern subspecies of red bat, *Lasiurus borealis borealis.*

Specimens examined (1). *Presidio Co.:* Sierra Vieja, ZH Canyon, 9 mi W Valentine, 1 (TTU).

References. 2, 4, 12, 15, 33, 80, 102, 182, 200, 218, 219, 265, 362.

Lasiurus seminolus (Rhoads, 1895)
Seminole Bat

Description. This is a medium-sized bat (forearm=35–45 mm) with short, rounded ears. As with other bats of the genus *Lasiurus*, the interfemoral membrane is densely furred dorsally. Pelage coloration is a dark reddish brown (or mahogany), and the hair tips are sometimes white, giving the fur a frosted appearance. *L. seminolus* may be confused with the red bat (*Lasiurus borealis*) from which it is distinguished as described in the account of the latter. Average external measurements of *L. seminolus* are: total length, 103 mm; tail, 44 mm; hind foot, 8 mm; ear, 12 mm.

Distribution. This bat is distributed across the southeastern United States, its distribution closely paralleling that of Spanish moss, the clumps of which serve as its roosting sites. It is a common resident of the Pineywoods region in eastern Texas where it reaches the westernmost limits of its range in Burleson County. *L. seminolus* has also been documented in the Gulf Prairies and Marshes region of Texas. The species is a year-round resident of Texas, although winter records are rare and none exist for the months of November and December. There are extralimital records, which are not substantiated by voucher specimens, from McLennan and Cameron Counties (see additional records).

Subspecies. *L. seminolus* (Rhoads, 1895) is a monotypic species with no subspecies recognized.

Life History. As mentioned above, this bat prefers to roost in clumps of Spanish moss, but it will also locate behind loose tree bark or in other foliage. It is a solitary bat and roosts are typically occupied by a single individual, or a female with young. The Seminole bat is thought to remain active throughout the year, although it begins to reduce activity when ambient temperatures reach 21°C (Constantine, 1958). These bats rarely fly when temperatures drop below 18°C. Specimens have been collected in East Texas from February through November (Schmidly et al., 1977).

Seminole bats feed over watercourses and clearings and generally at treetop level. Their flight is swift and direct, and they may occasionally alight on vegetation to capture prey. Specific prey items include true bugs, flies, beetles, and even ground-dwelling crickets. Seminole bats also frequent street lights to feed on insects attracted to the light.

Females may carry from one to four embryos but usually give birth to two young (Davis, 1974). In Texas, parturition occurs in late May or June. The young bats mature quickly and are able to fly at three to four weeks of age.

Out of a total of 186 specimens of Seminole bats reported to the Texas Department of Health, only 17 (9%) tested positive for the rabies virus.

Specimens examined (100). *Rusk Co.*: 6 mi E Mt. Enterprise, 1 (SFASU). *Panola Co.*: 7.5 mi ENE Carthage, 1 (TCWC). *Cherokee Co.*: 3 mi W Forest, 1 (SFASU). *Shelby Co.*: 4 mi S Joaquin, 1 (SFASU); 2.3 mi SE Patroon, 1 (TCWC). *Nacogdoches*

Lasiurus seminolus. Seminole Bat. (John L. Tveten)

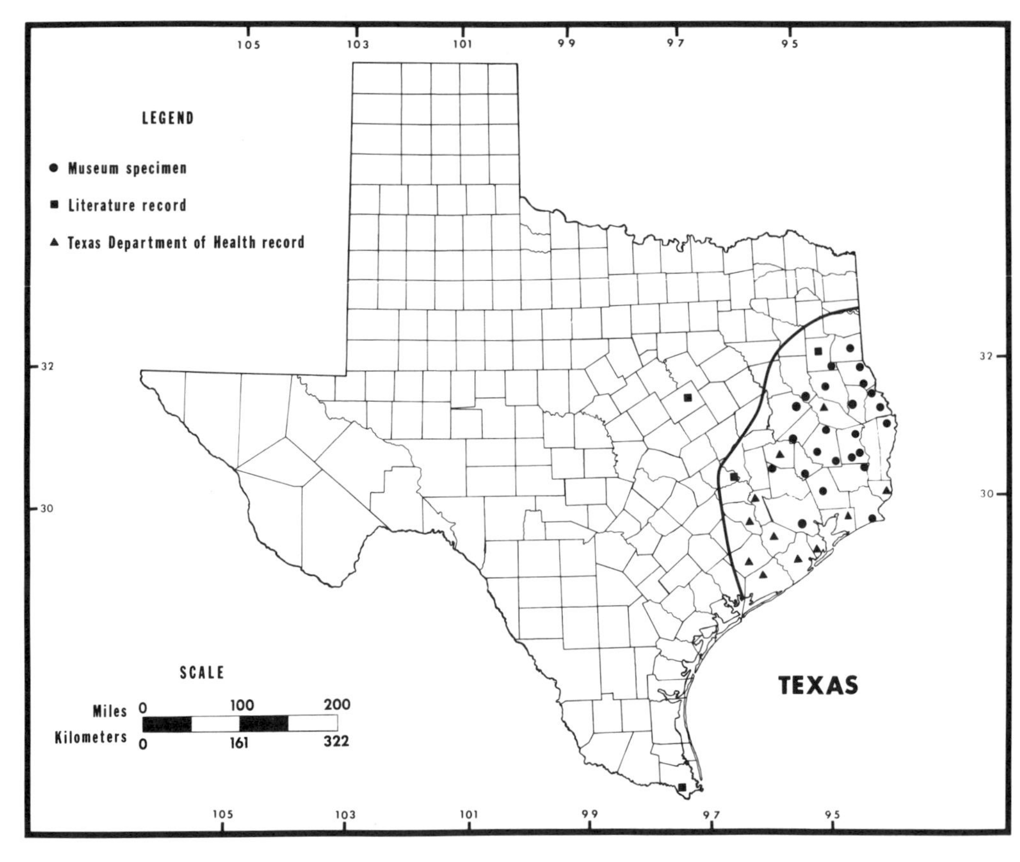

Map 15. Distribution of the Seminole bat, *Lasiurus seminolus.*

Co.: Nacogdoches, 6 (SFASU); SFA Campus, Nacogdoches, 3 (2 SFASU, 1 TTU); 9.5 mi SW Nacogdoches, 1 (TNHC). *San Augustine Co.*: near Broaddus, 1 (FWMSH). *Sabine Co.*: 5 mi N Geneva, 1 (TCWC); Hemphill, 1 (SFASU). *Houston Co.*: 6.3 mi N Ratcliff, 1 (UTACV). *Trinity Co.*: 3 mi E Trinity, 1 (TCWC). *Newton Co.*: 11.5 mi N Burkeville, 6 (TCWC); 8.5 N Burkeville, 1 (TCWC). *Polk Co.*: 1 mi S Neches River, 3 mi E Hwy 59, 2 (SFASU); 4 mi E Livingston, 1 (TCWC); 3 mi S Livingston, 4 (TCWC); 2 mi NNW Segno, 2 (TCWC); 2 mi E, 1.7 mi S Camp Ruby, 2 (TCWC). *Tyler Co.*: Beech Creek Unit, BTNP, 1.1 mi W, 1 mi S Town Bluff, 1 (TCWC); 3.6 mi S, 2.9 mi W Town Bluff, 8 (TCWC); 2 mi S, 1.5 mi W Town Bluff, 2 (TCWC); 2.8 mi S, 2.8 mi W Town Bluff, BTNP, 1 (TCWC); 3.8 mi N, 1.9 mi W Spurger, BTNP, 2 (TCWC); 0.6 mi N, 0.7 mi W Spurger, 2 (TCWC); Turkey Creek Unit, BTNP, 4.3 mi S, 4.5 mi E Warren, 1 (TCWC). *San Jacinto Co.*: Farm Rd. 945 on San Jacinto River, 10 (TTU); 5 mi NW Cleveland, 15 (TTU). *Montgomery Co.*: 5 mi E Richards, 1 (TCWC). *Hardin Co.*: 8.6 mi N, 3.8 mi E Silsbee, 1 (TCWC). *Liberty Co.*: 12 mi N Dayton, 2 (TTU). *Jefferson Co.*: 11.7 mi W Sabine Pass, TX Hwy. 87, 1 (TCWC). *Harris Co.*: Houston, 15 (2 LACM, 8 LSUMZ, 4 TCWC, 1 WMM).

Additional records: *Rusk Co.*: no specific locality (McCarley, 1959). *McLennan*

Co.: Waco (Strecker, 1924). *Burleson Co.*: 6 mi N Clay (Lee, 1987). *Cameron Co.*: Brownsville (Hall, 1981).

References. 4, 7, 12, 13, 20, 21, 25, 27, 33, 61, 76, 80, 82, 89, 113, 132, 265, 283, 294, 339.

Lasiurus cinereus (Palisot de Beauvois, 1796)
Hoary Bat

Description. This is a large (forearm=46–58 mm, weight=20–35 g) and heavily furred bat with short, rounded ears and a striking pelage coloration. Overall, the fur is a dark mahogany brown but the hair tips are white, giving the bat a distinctive frosted appearance. Fur on the throat and face is yellowish, and the dorsal surface of the interfemoral membrane is densely furred, as in other *Lasiurus* species.

Due to its large size and beautiful, distinctive fur the hoary bat is not easily confused with any other bat. In flight, this bat produces a chattering noise similar to that of the big brown bat (*Eptesicus fuscus*). Average external measurements are: total length, 131 mm; tail, 59 mm; hind foot, 12 mm; ear, 17 mm.

Distribution. This is a forest-dwelling, transcontinental species which has been recorded from scattered localities throughout Texas. It has been reported from all ecological regions in the state. In the Trans-Pecos, it is restricted to mountainous, wooded areas.

Hoary bats are migratory and the sexes appear to segregate geographically in summer. Females are known to migrate through Texas in spring and fall; males first appear in spring and remain throughout the summer in small numbers. Although the overall pattern for this species in Texas is one of a spring-fall migrant, hoary bats may also overwinter in the state.

Subspecies. Populations of this bat in the continental United States, Mexico, and Guatemala are referrable to the subspecies *L. c. cinereus* (Palisot de Beauvois, 1796). Different subspecies occur in Hawaii and South America (Hall, 1981).

Life History. The hoary bat typically roosts singly in tree foliage 3.0–4.6 m above the ground and often at the edge of a clearing. These bats are migratory and exhibit an interesting seasonal distribution as a result. In summer, females move to the northern, eastern, and central United States to give birth and raise their young. Males, however, remain in the western states, generally in montane areas. This pattern of migration and sexual segregation is illustrated in Texas by the distribution of this species in the Chisos Mountains of Big Bend National Park. Here, hoary bats are a rare summer resident and only males occur in the mountains at this time. In the spring and fall, however, only females are found (Easterla, 1973).

Interestingly, females are believed to precede males in migratory movements, which contrasts with the pattern found in migratory birds. Hoary bats breed on

Lasiurus cinereus. Hoary Bat. (John L. Tveten)

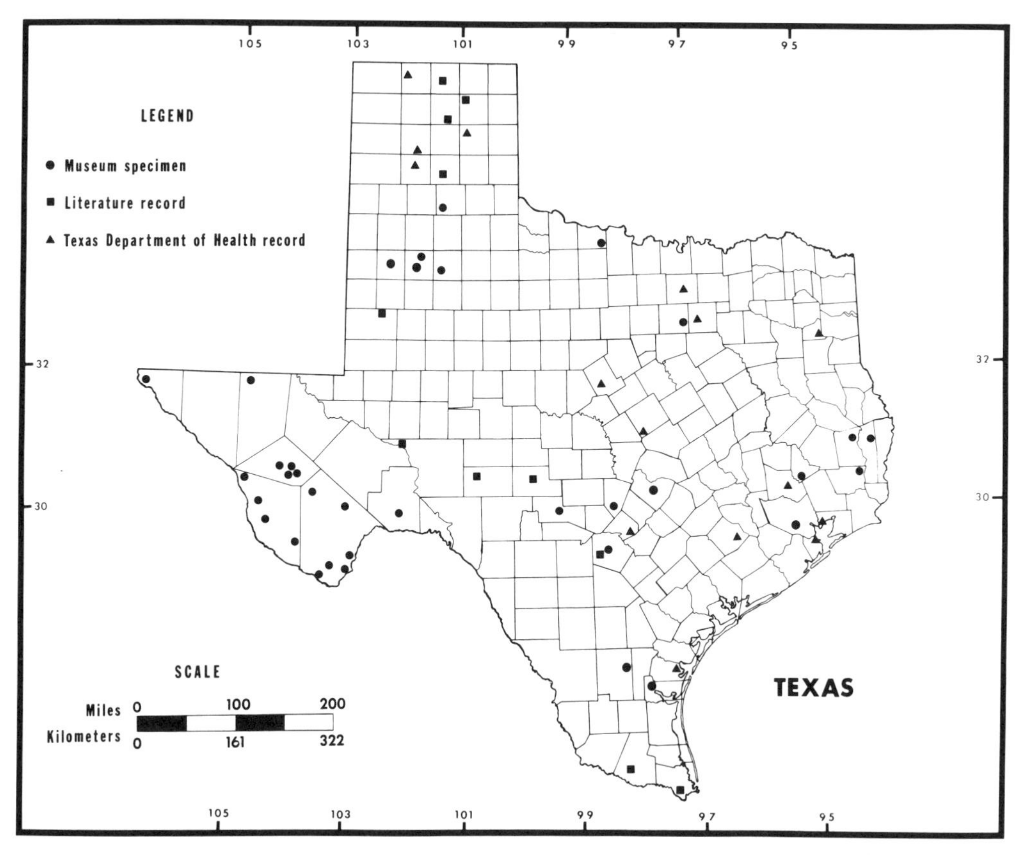

Map 16. Distribution of the hoary bat, *Lasiurus cinereus.*

the winter range prior to migration, whereas birds breed on summer range (Findley and Jones, 1964). Unlike the situation in migratory birds, male hoary bats have no need to set up territories or otherwise prepare for breeding prior to the female's arrival and simply do not make the long journey north.

Parturition seems to range from mid-May into early July; the usual number of young per litter is two and ranges from one to four (Shump and Shump, 1982). Parturition in Texas has been documented during May and June.

Hoary bats are powerful fliers and emerge late in the evening to forage. They forage on the wing, cover large areas, and also forage considerable distances from diurnal roost sites (Black, 1974). They seem to have a strong preference for moths (Black, 1972; Ross, 1967). Indications are that the bat approaches a flying moth from the rear, engulfs the prey's abdomen-thorax in its mouth and then bites down, allowing the sheared head and wings to drop to the ground (Shump and Shump, 1982). Hoary bats are also known to eat beetles, flies, grasshoppers, termites, dragonflies, and wasps. Known predators include rat snakes.

Forty hoary bats have been reported to the Texas Department of Health and ten of these proved rabid. This is the highest incidence (25%) of rabies for any Texas bat, but the sample is too small to warrant any definitive conclusions regarding the species' susceptibility to the rabies virus.

Specimens examined (93). *Briscoe Co.*: Los Lingos Canyon, 1 (TTU). *Wichita Co.*: Wichita Falls, 1 (MWSU). *Hockley Co.*: Pep, 1 (UIMNH). *Lubbock Co.*: Lubbock, 5 (TTU); 3 mi E New Deal, 1 (TTU). *Crosby Co.*: 13.5 mi S, 2.1 mi W Ralls, 1 (TTU). *Tarrant Co.*: Arlington, 1 (UTACV). *El Paso Co.*: Yarborough Dr., Lower Valley, El Paso, 1 (UTEP). *Culberson Co.*: The Bowl, GMNP, 9 (TTU); McKittrick Canyon, GMNP, 12 (TCWC); 2 mi NW Pine Springs, GMNP, 2 (TTU). *Jasper Co.*: 1.0 mi E Jasper, Sherwood Forest Ranch, 2 (TCWC). *Tyler Co.*: 3.6 mi S, 2.9 mi W Town Bluff, 2 (TCWC). *Jeff Davis Co.*: Reynolds Ranch near Rockpile Park, 1 (ASVRC); Davis Mts., 1 (USNM/ FWS); 8 mi S jct. hwys. 166 and 118, 4 (2 SRSU, 2 TTU); 5 mi E Mt. Livermore, 1 (UMMZ); Limpia Canyon, 5 mi N Ft. Davis, 1 (TCWC); 3.5 mi NE Ft. Davis, 1 (TTU); Fort Davis, 1 (UMMZ); 2 mi W Ft. Davis, 8 (UMMZ); Old McGuire Homestead, Davis Mts. Resort, 1 (SRSU); DMSP, 4 (TTU); 1.5 mi E DMSP, 1 (TTU). *San Jacinto Co.*: 5 mi NW Cleveland, 1 (TTU). *Terrell Co.*: 6 mi S Dryden, 1 (MWSU). *Presidio Co.*: ZH Canyon, Sierra Vieja, 2 (TTU); Sierra Vieja, 9 mi W Valentine, 1 (SRSU); 8 mi NE Candelaria, 1 (TCWC); Fred Shely Ranch, Chinati Mts., 1 (SRSU); Pinto Canyon, Chinati Mts., 14 mi E Ruidosa, 3 (TTU); 5 mi SE Bandera Mesa, 2 (MWSU). *Brewster Co.*: Alpine, 1 (SRSU); 14 mi E Marathon, 1 (TCWC); Old Ore Rd., BBNP, 1 (SRSU); Rio Grande Village, BBNP, 2 (BBNHA); Basin Sewage Lagoon, BBNP, 1 (BBNHA); Mt. Emory, Chisos Mts., 7,100 ft., BBNP, 1 (FMNH); Hot Springs, 1 (UMMZ); Big Bend of the Rio Grande, 2,000 ft., 1 (UMMZ). *Travis Co.*: Austin, 1 (TCWC). *Hardin Co.*: 10.9 mi N, 2.3 mi E Silsbee, BTNP, 1 (TCWC). *Blanco Co.*: Blanco, 1 (TCWC). *Kerr Co.*: Hughes Ranch, 1 (WMM); no specific locality, 1 (WMM). *Bexar Co.*: San Antonio, 1 (TCWC). *Duval Co.*: San Diego, 1 (ANSP). *Kleberg Co.*: Kingsville, 1 (TAIU); 7.5 mi S Kingsville, 1 (TAIU).

Additional records: *Hansford Co.*: Gruver (Cutter, 1959). *Hutchinson Co.*: Borger (Wiseman et al., 1962). *Roberts Co.*: 10 mi S, 15 mi E Spearman (Jones et al., 1988). *Armstrong Co.*: 10 mi NE Wayside (Eads et al., 1957). *Tarrant Co.*: Arlington (Wilkins et al., 1979). *Gaines Co.*: Seagraves (Eads et al., 1957). *McLennan Co.*: no specific locality (Davis, 1974). *Jeff Davis Co.*: 1 mi E McDonald Observatory (Baker and Patton, 1967). *Crockett Co.*: 5 mi S, 5 mi E McCamey (Manning et al., 1987). *Kimble Co.*: Texas Tech Univ. Center at Junction (Manning et al., 1987); 5 mi S Texas Tech Univ. Center at Junction (Manning et al., 1987). *Sutton Co.*: 13 mi W Sonora (Manning et al., 1987). *Bexar Co.*: Cubbra Springs, 18 mi W San Antonio (Allen, 1922). *Hidalgo Co.*: Edinburg (Muliak, 1943). *Cameron Co.*: no specific locality (Davis, 1974).

References. 1, 2, 3, 4, 6, 7, 10, 12, 13, 15, 20, 21, 22, 24, 25, 26, 27, 29, 31, 33, 53, 68, 71, 76, 80, 82, 84, 89, 91, 92, 93, 96, 97, 99, 102, 105, 113, 141, 142, 145, 151, 155, 167, 169, 181, 188, 193, 198, 201, 250, 256, 257, 262, 265, 267, 268, 279, 288, 293, 304, 314, 339, 342, 356, 362, 363, 389, 414, 418, 424.

Lasiurus intermedius H. Allen, 1862
Northern Yellow Bat

Description. This is a large bat (forearm = 45–52 mm) with short, pointed ears and long wings. Pelage coloration is yellowish brown to yellowish gray and only

Lasiurus intermedius. Northern Yellow Bat. (John L. Tveten)

the anterior half of the interfemoral membrane is furred. This bat is easily distinguished from other members of the genus *Lasiurus* by its pointed, rather than rounded, ears and yellowish pelage coloration.

The northern yellow bat may be confused with the southern yellow bat (*Lasiurus ega*). These two bats, which may occur together in extreme South Texas, can be separated on the basis of size, with *L. ega* being much smaller than *L. intermedius.* Average external measurements of *L. intermedius* are: total length, 133 mm; tail, 55 mm; hind foot, 10 mm; ear, 17 mm.

Distribution. The northern yellow bat occurs primarily along the Gulf Coast in the Gulf Prairies and Marshes region from Harris County south to Cameron County, and also the South Texas Plains. Inland records are uncommon but include specimens from the Pineywoods, Post Oak Savannah, Blackland Prairies, and Edwards Plateau regions.

Although a spring, summer, and fall resident, the northern yellow bat is relatively rare in Texas. It has not been found in great numbers at any locality in the state.

Subspecies. Texas specimens are referrable to two subspecies, according to the latest evaluation of the species' taxonomy (Hall and Jones, 1961). *L. i. floridanus* (G. S. Miller, 1902) is known from Bexar and Travis Counties eastward and *L. i. intermedius* H. Allen, 1862, from San Patricio County southward. *L. i. floridanus* is generally smaller and slightly less yellow in pelage coloration than *L. i. intermedius.*

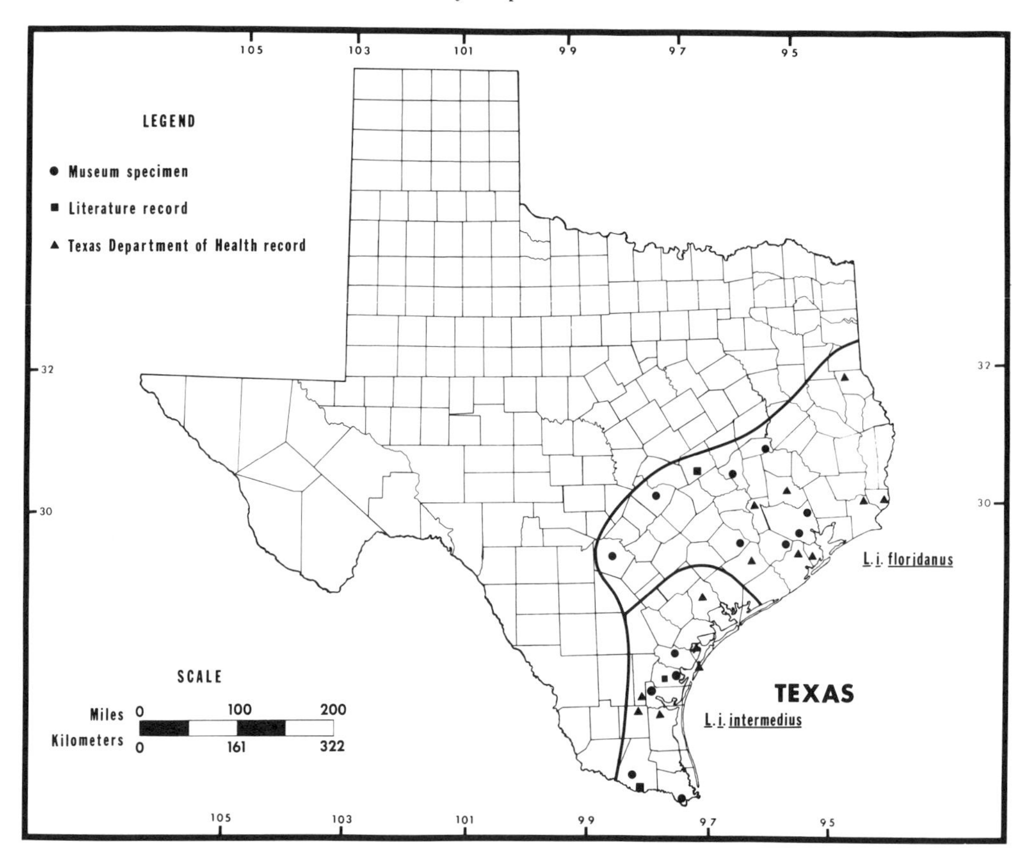

Map 17. Distribution of the two subspecies of the northern yellow bat, *Lasiurus intermedius*.

Life History. The distribution of this bat in the United States closely coincides with that of Spanish moss, which is its preferred roosting site. In South Texas, however, *L. intermedius* roost in palm trees where individual bats are well concealed behind the large fronds. A single roosting site may contain several bats and such groups are often quite noisy, especially when young are present; their bickering quickly gives them away (Davis, 1974). Migration and winter habits are poorly known. *L. i. floridanus* has been reported in Texas in late winter, but *L. i. intermedius* has not been recorded during this season.

Northern yellow bats prefer to forage over open, grassy areas such as pastures, lake edges, golf courses, and along forest edges. In Florida, they often form groups or "feeding aggregations" when foraging. Such feeding aggregations are segregated by sex; males are rarely found in such groups, and they seem to be more solitary in their habits than females. Specific prey items include leafhoppers, dragonflies, flies, diving beetles, beetles, ants, and mosquitoes.

Females carry three to four embryos in spring and litter size is believed usually to be two or three. Parturition probably occurs in late May or June in Texas.

Out of 126 northern yellow bats reported to the Texas Department of Health, 11 (8.7%) tested positive for rabies.

Specimens examined. *Lasiurus intermedius floridanus* (10): *Madison Co.*: Bedias Creek, Madison Co. and Walker Co. line, 1 (SFASU). *Brazos Co.*: 8 mi SW College Station, 300 ft., 1 (TCWC). *Travis Co.*: Austin, 1 (TNHC). *Harris Co.*: 4 mi W Huffman, 1 (TCWC); Houston, 3 (1 KU, 2 TCWC). *Colorado Co.*: Eagle Lake, 1 (TCWC). *Fort Bend Co.*: Brawner Ranch, 2 mi S Dewalt, 1 (TTU). *Bexar Co.*: San Antonio, 1 (TCWC).

Additional records: *Milam Co.*: Rockdale (Wiseman et al., 1962). *Harris Co.*: Houston (Wiseman et al., 1962).

Lasiurus intermedius intermedius (86). *San Patricio Co.*: Sinton, 2 (TNHC). *Nueces Co.*: Corpus Christi, 5 (CCSU). *Kleberg Co.*: Kingsville, 15 (14 TAIU, 1 TCWC). *Hidalgo Co.*: 5.6 mi N Mission, 2 (TCWC); 5.4 mi N Mission, 1 (TCWC). *Cameron Co.*: Harlingen, 2 (1 LACM, 1 MVZ); Brownsville, 54 (4 AMNH, 3 FMNH, 1 ANSP, 2 TCWC, 44 USNM/FWS); 5 mi SE Brownsville, 5 (TTU).

Additional records: *Nueces Co.*: 1 mi S Driscoll (Spencer et al., 1987). *Hidalgo Co.*: Santa Ana Wildlife Refuge (Hall and Jones, 1961).

References. 1, 4, 7, 12, 20, 21, 25, 29, 33, 60, 77, 80, 82, 84, 89, 96, 113, 132, 170, 215, 218, 265, 294, 362, 363, 378.

Lasiurus ega (Gervais, 1856)
Southern Yellow Bat

Description. This is a medium-sized bat (forearm=45–48 mm) with relatively long wings and short, pointed ears. The dorsal surface of the interfemoral membrane is incompletely furred with only the anterior half covered by hair. Pelage coloration is yellowish brown.

The southern yellow bat may be confused with its northern relative (*Lasiurus intermedius*) and identification may be difficult in South Texas where the two species are sympatric. Distinguishing features are listed in the account of *L. intermedius.* Average external measurements for *L. ega* are: total length, 115 mm; tail, 44 mm; hind foot, 8 mm; ear, 13 mm.

Distribution. *L. ega* is a neotropical bat which has been recorded in the United States from southern California, southern Arizona, extreme southwestern New Mexico, and South Texas. This bat may be expanding its range in the United States, because of the increased usage of ornamental palm trees in landscaping. Palms are the preferred roosting site for this species in Texas (Spencer et al., 1988).

Most Texas specimens have been collected along the Rio Grande near Brownsville, where this bat is known to inhabit a natural grove of palm trees. The northernmost record in the state is from Corpus Christi in Nueces County (Spencer et al., 1988). This species appears to be a year-round resident of the Brownsville area, where it has been collected in six different months, including December (Baker et al., 1971).

Subspecies. Texas specimens are referrable to the subspecies *L. e. panamensis* (Thomas, 1901), according to Baker et al. (1988).

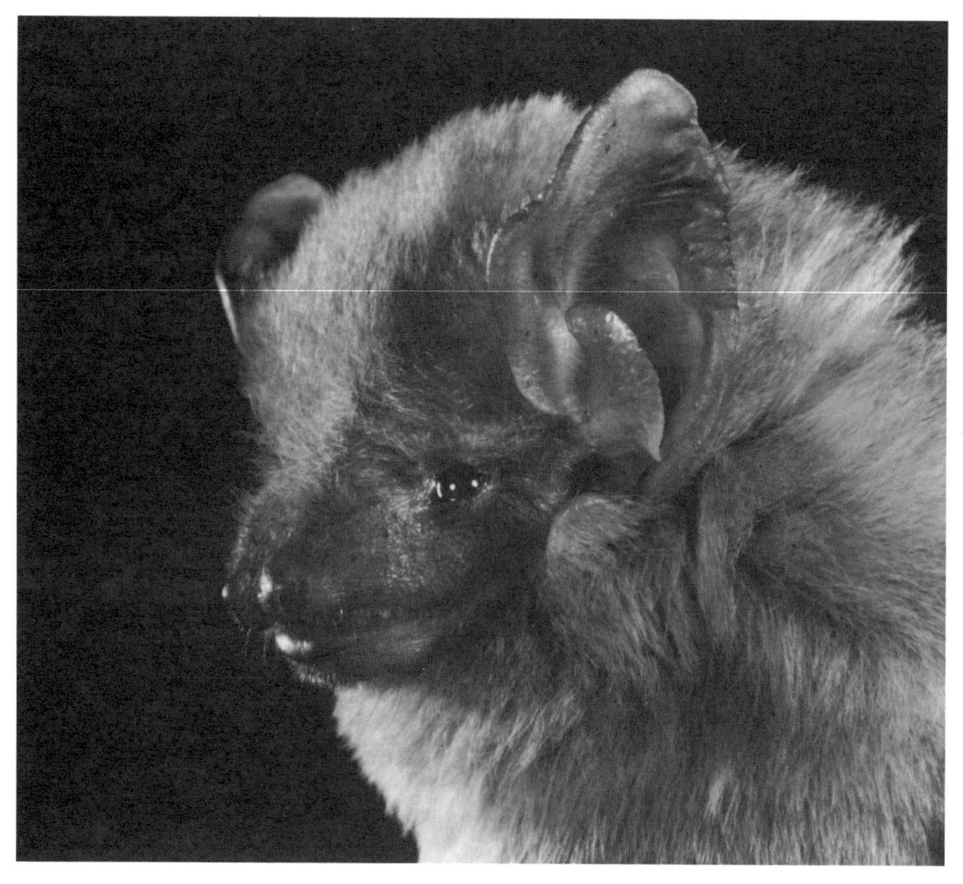

Lasiurus ega. Southern Yellow Bat. (Merlin D. Tuttle, Bat Conservation International)

Life History. This is one of the rarest and most restricted bats in Texas and very little is known about its biology. Apparently, it is a tree-roosting species and, as mentioned above, commonly roosts in palm trees. It may be migratory in parts of its range, but it appears to be a permanent resident of Texas (Baker et al., 1971).

Its food habits are not well documented although small to medium-sized, night-flying insects are probably the main prey items. Reproductive data are equally scarce. In Texas, *L. ega* is believed to breed in late winter, since pregnant females have been collected in April and June (Baker et al., 1971). These individuals carried two, three, or four embryos, suggesting that litter size may be larger than in most other species of bats.

One specimen of *L. ega* has been reported to the Texas Department of Health and that individual tested negative for the rabies virus.

Remarks. Specimens from Texas were formerly assigned to the subspecies *L. e. xanthinus.* In genetic studies of *Lasiurus,* however, Baker et al. (1988) have shown that this subspecies should be elevated to specific status (*Lasiurus xanthinus*) and that *L. ega* from Texas probably belong to the subspecies *L. e. panamensis.*

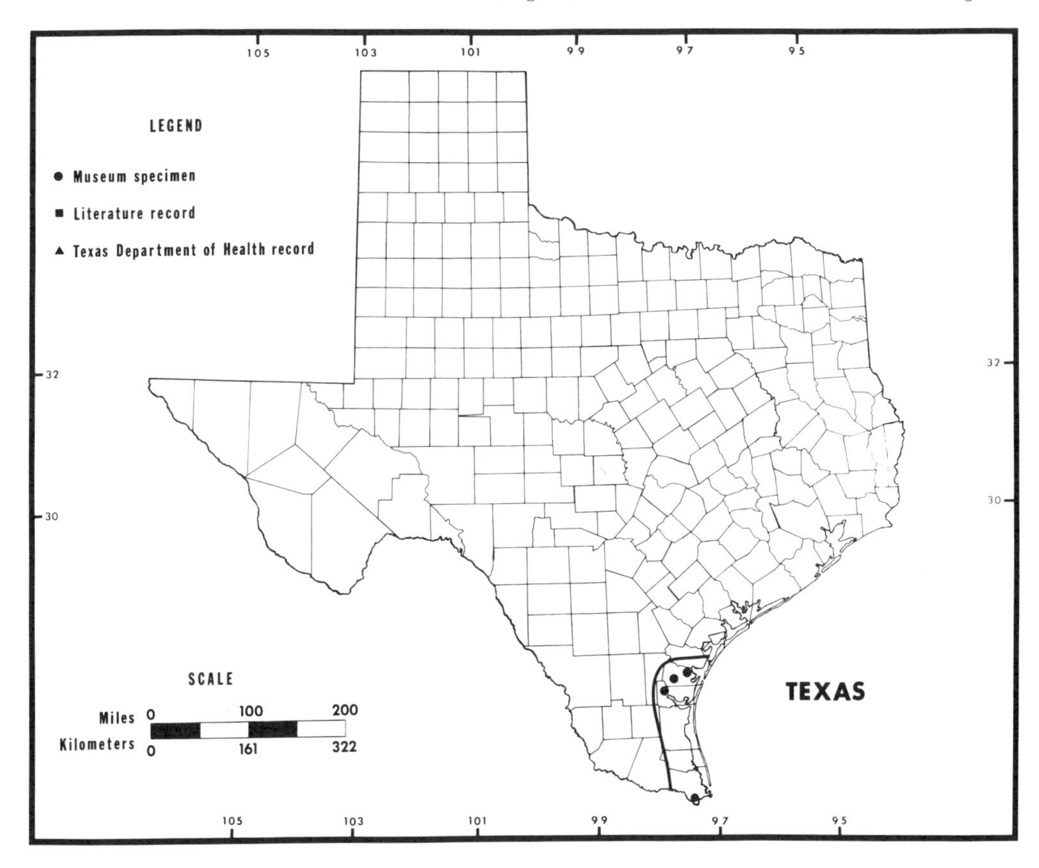

Map 18. Distribution of the southern yellow bat, *Lasiurus ega panamensis.*

Specimens examined (23). *Nueces Co.:* Corpus Christi, 1 (CCSU); 1 mi S Driscoll, 4 (CCSU). *Kleberg Co.:* Texas A&I Univ. campus, Kingsville, 1 (TAIU). *Cameron Co.:* Brownsville, 1 (AMNH); 5 mi SE Brownsville, 16 (1 TAIU, 15 TTU).

References. 1, 3, 4, 7, 10, 12, 15, 21, 29, 33, 80, 82, 83, 84, 89, 113, 198, 215, 218, 265, 356, 362, 378.

Nycticeius humeralis (Rafinesque, 1818)
Evening Bat

Description. This is a small (forearm=33–39 mm, weight=5–7 g), rather nondescript bat whose pelage coloration is dark brown dorsally and paler below. The wings are short and narrow and the ears are small. The wing and tail membranes are dark and leathery in texture, as are the ears. The calcar is not keeled and juvenile specimens are darker than adults. As with many vespertilionids, females typically are slightly larger than males.

N. humeralis resembles many species of *Myotis*, but may easily be distinguished by its short, blunt tragus—as opposed to the long and sharp-pointed tragus of

Nycticeius humeralis. Evening Bat. (Merlin D. Tuttle, Bat Conservation International)

myotises. Average external measurements for *N. humeralis* are: total length, 87 mm; tail, 32 mm; hind foot, 8 mm; ear, 12 mm.

Distribution. The evening bat is commonly encountered in seven ecological regions in the eastern half of Texas—the Cross Timbers and Prairies, Edwards Plateau, South Texas Plains, Blackland Prairies, Post Oak Savannah, Gulf Prairies and Marshes, and Pineywoods. The westward limits of its range stretch from Palo Pinto County in the north to Bandera, Kerr, and San Saba counties in central Texas, and southwestward to Kinney and Real counties.

The evening bat is a forest inhabitant and is commonly found along watercourses. It is most abundant in the eastern portion of its range in Texas and is probably the most common bat of the Pineywoods, where it occurs in all major vegetation zones and in all seasons of the year (Schmidly et al., 1977).

N. humeralis is migratory throughout much of its range in the United States and favors southern climes in winter. In Texas, the evening bat has been collected in abundance from late March through September but only occasionally in winter.

Subspecies. Texas specimens are referrable to the subspecies *N. h. humeralis* (Rafi-

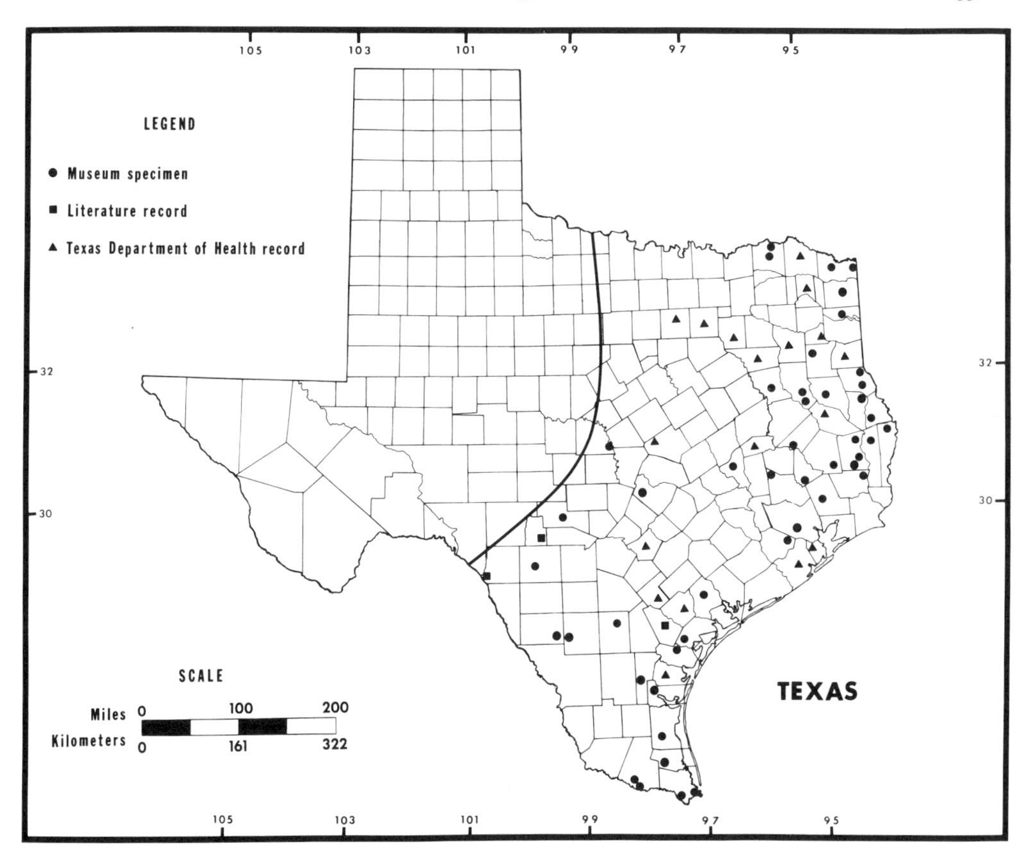

Map 19. Distribution of the evening bat, *Nycticeius humeralis humeralis.*

nesque, 1818), according to the latest taxonomic assessment of the species in the state (Schmidly and Hendricks, 1984).

Life History. The evening bat has been collected in all months of the year in Texas, indicating that it is a year-round resident of the state, although the winter habits of this bat are poorly known. Winter roosting sites have not been located, but in summer these bats commonly roost in hollow trees, behind loose tree bark, and in attics. One such roosting site, located in an apartment building in College Station, Brazos County, was reported to me by the owner in late June. This roost contained several hundred adult females and young bats. The temperature in the attic was extremely warm (about 45°C) and the bats were spaced evenly over the ceiling as well as in the wall spaces. The bats entered and left the attic along an open crack between the roof and the wall. They remained at this location until late August, at which time they dispersed and left the roost. Although I am not certain where these bats went, another colony of adult male and female bats was subsequently located in a hollow tree in the wooded area adjacent to the apartments.

Evening bats seem to have two preferred times of foraging, one in the early evening hours and then again just before dawn. Specific prey items include small, night-flying insects such as bugs, flying ants, spittle bugs, June beetles, pomace flies, Japanese beetles, and moths (Ross, 1967; Mumford and Whitaker, 1982). Predators of evening bats include black rat snakes, raccoons, and domestic and feral cats (Watkins, 1972).

Copulation takes place in the fall, but it is not known where this occurs (Watkins, 1972). Two young are born to the female in late May to early June. Nursery colonies may contain several hundred individuals and at this time the colonies are usually segregated by sex, with adult males rarely encountered in the nursery colonies. The young bats are volant at approximately twenty days and are nearly adult size by one month of age.

Of 221 evening bats reported to the Texas Department of Health, only six tested positive for rabies (a 3 percent infection rate).

Specimens examined (403). *Lamar Co.*: Arthur City, 1 (USNM/FWS); Paris, 3 (USNM/FWS). *Red River Co.*: no specific locality, 1 (DMNHT). *Bowie Co.*: 8 mi N New Boston, 1 (TCWC); Texarkana, 3 (USNM/FWS). *Marion Co.*: Jefferson, 1 (USNM/FWS). *Rusk Co.*: 1.6 mi NE New London, 6 (SFASU). *Cherokee Co.*: 1 mi E Alto, 1 (SFASU); 3 mi W Forest, 3 (SFASU). *Anderson Co.*: Palestine, Gus Engling Wildlife Mgt. Area, 1 (TCWC). *Shelby Co.*: 4 mi S Joaquin, 1 (SFASU); Shelbyville, 2 (SFASU). *Nacogdoches Co.*: 1 mi N Nacogdoches, 1 (SFASU); Nacogdoches, 91 (2 MSU, 56 SFASU, 33 TTU); Nacogdoches, SFA campus, 32 (1 DMNHT, 31 SFASU); 0.5 mi E Nacogdoches, 1 (SFASU); 1 mi E Nacogdoches, 1 (SFASU); 11 mi SW Nacogdoches, 1 (TTU). *San Augustine Co.*: 2.3 mi SE Patroon, 1 (SFASU). *Sabine Co.*: Pineland, 1 (TCWC). *San Saba Co.*: Gorman Falls Fish Camp, 4 (TCWC). *Trinity Co.*: 4 mi W Trinity, 1 (TCWC); Trinity, 6 (3 TCWC, 3 USNM/FWS); 4 mi E Trinity, 1 (TCWC). *Newton Co.*: 12 mi N Burkeville, 1 (TCWC); 11.5 mi N Burkeville, 2 (TCWC); 9.3 mi N Burkeville, 1 (TCWC); 9.2 mi N Burkeville, 2 (TCWC); 7.5 mi N Burkeville, 1 (TCWC). *Jasper Co.*: Jasper, 1 (USNM/FWS). *Polk Co.*: 5.5 mi W Dallardsville, 2 (TCWC); 4 mi W, 0.3 mi S Dallardsville, 4 (TCWC). *Tyler Co.*: 1.1 mi S, 1 mi W Town Bluff, 1 (TCWC); 3.6 mi S, 2.9 mi W Town Bluff, 8 (TCWC); Beech Creek Unit, BTNP, 0.6 mi N, 0.7 mi W Spurger, 9 (TCWC); 3.8 mi N, 1.9 mi W Spurger, BTNP, 3 (TCWC); 2.6 mi S, 3.9 mi E Warren, BTNP, 1 (TCWC). *Brazos Co.*: Bryan, 1 (TCWC); College Station, 5 (TCWC); 2 mi SE Bryan, 1 (TCWC); 3 mi S Bryan, 1 (TCWC); 8 mi SE College Station, 1 (TCWC). *San Jacinto Co.*: 5 mi NW Cleveland, 15 (TTU); jct. farm rd. 945 and E San Jacinto River, 10 (TTU). *Montgomery Co.*: 5 mi E Richards, 1 (TCWC). *Travis Co.*: 20 mi NW Bee Caves, by road, 1 (TCWC). *Hardin Co.*: 11 mi N, 2.3 mi E Silsbee, 1 (TCWC). *Liberty Co.*: 12 mi N Dayton, 2 (TTU). *Kerr Co.*: 5 mi SW Hunt, 3 (TCWC); 20 mi SW Hunt, 9 (1 KU, 8 TCWC). *Harris Co.*: Houston, 2 (1 USNM/FWS, 1 WMM). *Fort Bend Co.*: Brawner Ranch, 3 mi S Dewalt, 1 (LSUMZ). *Uvalde Co.*: 25 mi NW Uvalde, off hwy. 55, Cal Newton Ranch, 2 (TCWC). *Victoria Co.*: Victoria, 2 (USNM/FWS). *Dimmit Co.*: 3 mi S, 13 mi E Catarina, 4 (TCWC). *LaSalle Co.*: 1 mi N, 6 mi W Artesia Wells, 8 (TCWC). *McMullen Co.*: 5 mi SE Tilden, 1 (LSUMZ). *Refugio Co.*: Woodsboro, 1 (TCWC). *San Patricio Co.*: 8 mi N Sinton, 1 (TCWC). *Jim Wells Co.*: 5 mi S, 7 mi W Alice, 5 (TCWC). *Kleberg Co.*: Kingsville, 2 (TAIU). *Kenedy Co.*: Rudolph, Norias Div.

King Ranch, 7 (TCWC); 8 mi E Rudolph, Norias Div. King Ranch, 1 (TCWC); 11 mi E Rudolph, Norias Div. King Ranch, 1 (TCWC); 6 mi ESE Rudolph, Norias Div. King Ranch, 5 (TCWC). *Hidalgo Co.*: Bentsen State Park, 5 (TCWC); 6 mi S McAllen, 1 (TNHC); Lomita Ranch, 6 mi W Hidalgo, 2 (USNM/FWS). *Willacy Co.*: 5.75 mi S Raymondville, hwy. 77, 6 (TCWC). *Cameron Co.*: Brownsville, 33 (1 AMNH, 1 TCWC, 31 USNM/ FWS); 5 mi SE Brownsville, 57 (6 TAIU, 51 TTU); 17 mi E Brownsville, 2 (TCWC).

Additional records: *Erath Co.*: jct. of Erath, Somervell, and Hood County lines (Garner and Bluntzer, 1975). *San Saba Co.*: 4 mi SSE Bend (Wilkins et al., 1979). *Real Co.*: Leakey (Manning et al., 1987). *Kinney Co.*: 19 mi SE Del Rio (Manning et al., 1987). *Bee Co.*: 8 mi N Beeville (Blair, 1952).

References. 1, 3, 4, 6, 7, 12, 13, 20, 21, 22, 25, 26, 27, 29, 31, 33, 58, 76, 84, 89, 96, 97, 113, 142, 160, 188, 198, 199, 262, 265, 288, 294, 314, 327, 339, 356, 362, 363, 418.

Euderma maculatum (J. A. Allen, 1891)
Spotted Bat

Description. This is a relatively large (forearm=48–51 mm, weight=16–20 g), strikingly distinctive bat. It has jet black fur dorsally with a conspicuous white spot present on each shoulder and on the rump. Also, a white patch of fur is located at the base of each ear. The ears are huge—longer than in any other North American bat. Spotted bats are strong, swift fliers and may be recognizable at night by their distinctive, high-pitched calls (Easterla, 1973). Average external measurements are: total length, 126 mm; tail, 51 mm; hind foot, 12 mm; ear, 47 mm.

Distribution. In spite of a fairly wide range in the United States, *E. maculatum* is one of the rarest and least known of American bats. Although it has been captured throughout the western United States, the species does not appear to be abundant at any one locality. It was first recorded in Texas at Big Bend National Park by David Easterla (1970a). Additional specimens have since been obtained from several localities within the boundaries of Big Bend National Park, but no specimens have been captured outside the park. The Big Bend area represents the southeasternmost locale where this bat has been recorded in the United States.

Its distribution is apparently unaffected by plant associations or elevation. Spotted bats have been captured in habitats ranging from ponderosa pine forests 2,450 m in elevation to semi-arid, desert scrub. The single habitat component seemingly consistent with the capture of spotted bats is the presence nearby of broken canyon country or cliffs, which are their preferred roosting sites (Easterla, 1973).

Subspecies. *E. maculatum* (J. A. Allen, 1891) is monotypic and no subspecies are recognized.

Life History. Winter habits of this species are poorly known. Easterla (1973) suggested that it is solitary except for possible hibernating clusters, but he did not

Euderma maculatum. Spotted Bat. (Merlin D. Tuttle, Bat Conservation International)

indicate whether spotted bats hibernated at Big Bend National Park. The species has been captured in Texas only during the months of June, July, and August. In summer, spotted bats take refuge in cracks and rock crevices of cliffs and rugged canyon country (Easterla, 1973; Poche, 1975; Poche and Ruffner, 1975).

Easterla (1973) examined the stomach contents of fifteen spotted bats from Big Bend National Park and found 97.1% moths, 2.7% June bettles, and 0.2% unidentified insects. Other studies (Easterla, 1965; Ross, 1961) have analyzed stomachs and found the contents to contain 100% small moths. In southeastern Utah, Poche and Bailie (1974) observed spotted bats feeding on small insects within 2 m of the ground and also witnessed this bat landing on the ground to capture insects. In captivity, *E. maculatum* has been induced to eat mealworms (Constantine, 1961a) and cottage cheese and flies (Handley, 1959). However, in the wild these bats are thought to feed almost exclusively on small moths.

Easterla (1973) captured one pregnant female at Big Bend National Park on June 11, 1969, and captured lactating females from June 11 to August 9. Thus, parturition probably occurs from late May to mid-June at Big Bend National Park. The scant evidence available indicates that only one young is born annually.

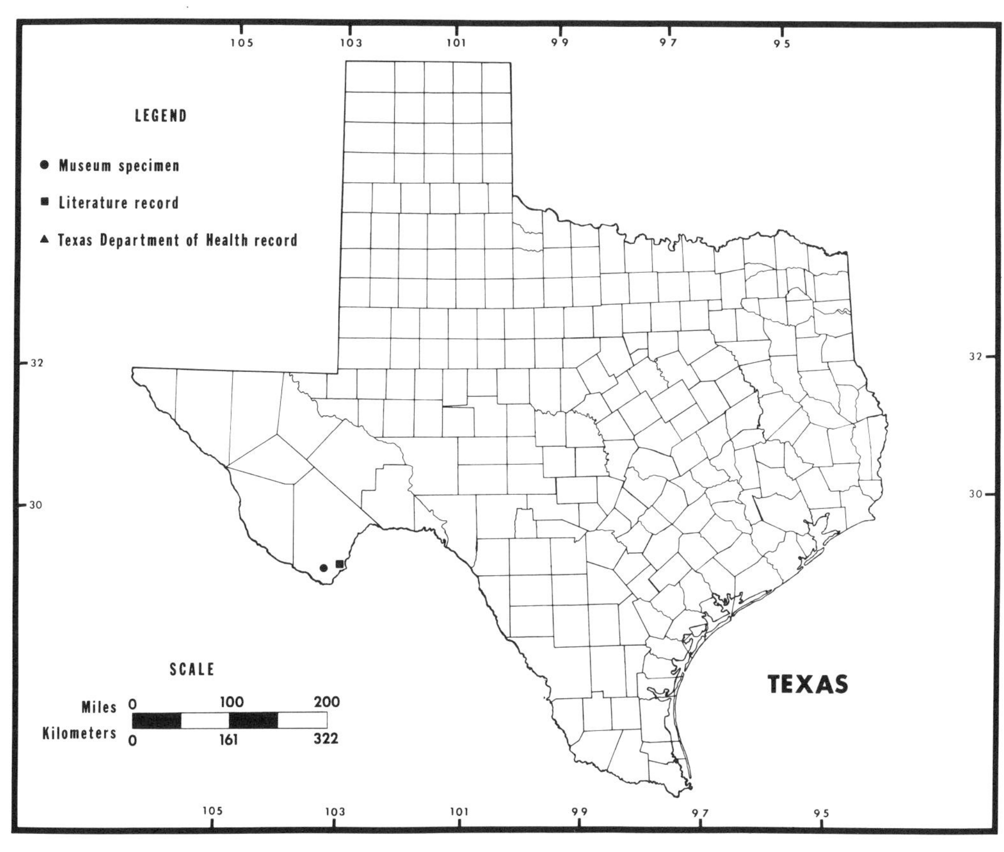

Map 20. Distribution of the spotted bat, *Euderma maculatum*.

Specimens examined (3). *Brewster Co.*: BBNP, 3 (1 FMNH, 1 LSUMZ, 1 TTU); over waterhole, BBNP, 2 (BBNHA).

Additional records: *Brewster Co.*: X-site, BBNP (Easterla, 1973); Ernst Canyon, BBNP (Easterla, 1973); Rio Grande Village *Gambusia* pools, BBNP (Easterla, 1973).

References. 4, 7, 10, 12, 15, 21, 24, 31, 33, 59, 92, 134, 174, 176, 178, 181, 183, 184, 185, 186, 198, 217, 256, 257, 265, 333, 334, 335, 355, 356, 414.

Plecotus townsendii Cooper, 1837
Townsend's Big-eared Bat

Description. This is a medium-sized bat (forearm=39–48 mm, weight=7–12 g) whose most distinctive features are its extremely large ears, which typically measure over 2.5 cm in length, and the presence of a large and distinctive facial gland on either side of the snout. The function of these glands is unclear, although they may secrete pheromones important in mating rituals. Pelage coloration in *P. townsendii* is uniformly pale to dark brown dorsally and ventrally. Females are typically slightly larger than males. Average external measurements are: total length, 98 mm; tail, 46 mm; hind foot, 11 mm; ear, 34 mm.

Plecotus townsendii. Townsend's Big-eared Bat. (John L. Tveten)

Townsend's big-eared bat is similar in appearance to the pallid bat (*Antrozous pallidus*) and to Rafinesque's big-eared bat (*Plecotus rafinesquii*). *A. pallidus* is much lighter in coloration—yellowish above with white below—and lacks the two nose lumps. *P. rafinesquii*, which also has the two nose glands, is found only in the extreme eastern portion of Texas and is not sympatric with *P. townsendii* in Texas. In addition, Rafinesque's big-eared bat has long toe hairs which extend beyond the tips of the claws, and it has a gray dorsum with light underparts.

Distribution. Townsend's big-eared bat ranges across the entire western United States (with a narrow range extension to Virginia) and is known in Texas from four ecological regions—the High Plains, Rolling Plains, Edwards Plateau, and Trans-Pecos. Its distribution is not restricted by vegetative associations. Specimens have been captured in habitats ranging from desert scrub to piñon-juniper woodlands. However, the presence of rocky, broken country is consistent with the capture of these bats. This is perhaps the most characteristic bat of caves and mine tunnels in the Trans-Pecos (Schmidly, 1977).

P. townsendii hibernates in caves across its range and is a year-round resident of Texas. It is one of the few species of Trans-Pecos bats that may be found regularly in winter, and records also exist from the High Plains and Rolling Plains during this season.

Subspecies. Texas specimens are referrable to the subspecies *P. t. pallescens* (G. S. Miller, 1897), according to the latest taxonomic revision of the species (Handley, 1959).

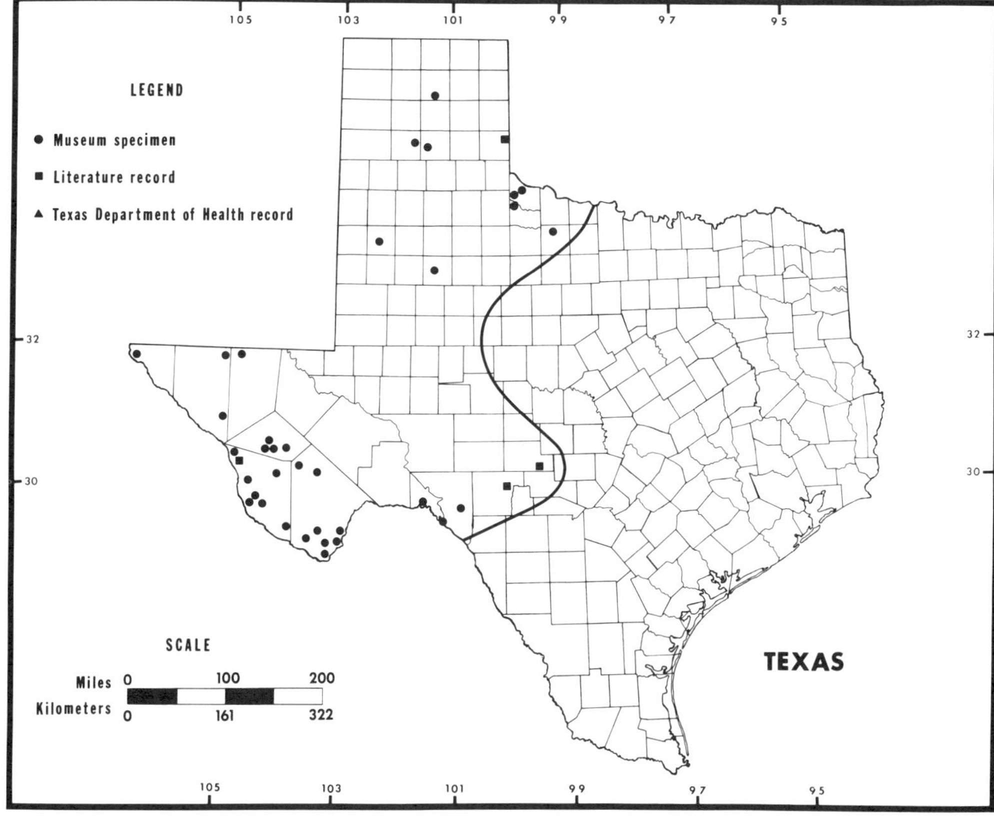

Map 21. Distribution of Townsend's big-eared bat, *Plecotus townsendii pallescens*.

Life History. *P. townsendii* is primarily a cave-dwelling species, but it occasionally occupies buildings. These bats inhabit gypsum caves found throughout northern Texas and are common in caves and mine tunnels of the Trans-Pecos. They do not use rock crevices and cracks as do many other species; instead, they typically hang by one foot from the roost ceiling. In a cave or mine these bats select roost sites with dim light near the zone of total darkness (Kunz and Martin, 1982), although I have found individuals on occasion hanging in fairly well-lit areas of abandoned buildings. The bats are intolerant of disturbance and will quickly abandon a roost site that has been disturbed.

P. townsendii hibernates throughout its range in caves and mines during winter months when temperatures are above freezing but below 12°C. The bats hibernate in tight clusters, which may help stabilize body temperature against external changes in temperature. While the animals are torpid, their large ears are coiled from the tips down and laid back against their necks. Sexual differences in winter activity and hibernacula site selection have been noted by Barbour and Davis (1969), who suggest that male bats select warmer hibernation sites and are more easily aroused and active compared with females.

Males and females occupy separate roosting sites during summer. During this season, males appear to lead a solitary life-style while females and young form maternity colonies that may number from 12 to 200 individuals, although in the eastern United States a colony of up to 1,000 males and females has been observed in Kentucky in the fall (Rippy and Harvey, 1965).

Reproduction in these bats has been studied extensively in California (Pearson et al., 1952) and, although timing in Texas populations may be somewhat different, general reproductive habits and behaviors are probably similar. Most breeding occurs at winter roosts from November to February, although some females may be inseminated prior to their arrival at the winter roosts. The sex organs of males do not develop fully until their second year of life, which means that young male bats are not reproductively active in their first year. Young females, however, may breed as early as four months of age. Males have been observed copulating with torpid females (Pearson et al., 1952), which is one reason why young females may become parous at such an early age. The large nose glands are probably of importance in precopulatory behavior. Males approach the females while making a twittering sound and vigorously rub the snout over the female's face, neck, forearms, and ventral surface, producing a strong and noticeable odor (Barbour and Davis, 1969).

A firm vaginal plug does not form in females following copulation, which means that repeated inseminations can occur during winter. This situation is unlike that of several other vespertilionids, where repeated copulations are prevented by the presence of such a plug. Sperm is stored by the female until ovulation occurs in late winter or early spring. Fertilization of the ovum takes place shortly afterward. Climatic factors and temperature may affect the length of gestation, which ranges from 56 to 100 days.

A single young is born in late May to early June. The baby bat weighs approximately 2.4 g at birth and is pink, naked, and completely helpless. At four days the newborn bat begins to display hair growth and by one month of age is volant and nearly adult size. At two months the juveniles are weaned and the nursery colonies begin to disperse.

These bats emerge late in the evening to forage and are swift, highly maneuverable fliers. Prey items include small moths, flies, lacewings, dung beetles, and sawflies (Ross, 1967).

Specimens examined (198). *Hutchinson Co.*: Borger, 1 (TTU). *Randall Co.*: cave in Palo Duro Canyon, 1 (TCWC). *Armstrong Co.*: 12 mi SSE Washburn, Palo Duro Canyon, 5 (TCWC); 29 mi SSW Claude, 4 (TTU). *Hardeman Co.*: 13 mi N Quanah, 1 (MWSU); 2 mi SE Lazare, 1 (MWSU); 3 mi S Lazare, 3 (UIMNH); Acme, 2 (MWSU); 9 mi W Quanah, 1 (MWSU); Quanah, 1 (MWSU); 7 mi E Quanah, 2 (MWSU); 9 mi SW Quanah, 4 (TTU); 10 mi SW Quanah, 1 (MWSU); 10 mi WSW Quanah, 5 (MWSU); 11 mi WSW Quanah, 19 (MWSU); 11 mi SW Quanah, 8 (MWSU); 12 mi WSW Quanah; 7 (MWSU); 12 mi SW Quanah, 1 (MWSU); 13 mi SW Quanah, 3 (MWSU); 15 mi SW Quanah, 1 (MWSU). *Foard Co.*: 12 mi WSW Lazare, 3 (MWSU); 20 mi SW Quanah, 1 (MWSU). *Hockley Co.*: Levelland, 1 (TTU). *Baylor Co.*: Lake Kemp, 1 (MWSU). *Garza Co.*: Post, 3 (TTU). *El Paso Co.*: Old Tin Mine, Franklin Mts., 1 (UTEP); El Paso, 1 (UTEP). *Hudspeth Co.*: Upper Sloth Cave, Guadalupe Mts., 1 (TCWC); Eagle Mt. Ranch, W Van Horn,

1 (SRSU). *Culberson Co.:* 7 mi N Pine Springs, GMNP, 1 (TCWC); stone cabin near Grisham Hunter Lodge, GMNP, 3 (TTU); McKittrick Canyon, GMNP, 5 (4 TCWC, 1 KU); The Bowl, GMNP, 1 (TTU); Lost Mine Peak, GMNP, 1 (TTU); Manzanita Spring, GMNP, 1 (TTU); Upper Dog Canyon, GMNP, 1 (TTU). *Jeff Davis Co.:* 8 mi S jct. hwys. 166 and 118, 1 (TTU); 3 mi E DMSP, 1 (TTU); 3.5 mi NE Ft. Davis, 1 (TTU); 3 mi NE Ft. Davis, 2 (TTU); 17 mi W Ft. Davis, 1 (TTU); 3 mi E jct. hwys. 166 and 505, 1 (TTU). *Presidio Co.:* 11 mi W Valentine, 3 (TNHC); 10 mi SW Valentine, 1 (TNHC); 8 mi NE Candelaria, 1 (TCWC); 8 mi S Marfa, 1 (SRSU); San Esteban Tunnel, 1 (SRSU); Pinto Canyon, Chinati Mts., 14 mi E Ruidosa 1 (UIMNH); Chinati Rancho, 23 mi NW Presidio, 1 (TCWC); Shafter Mine, Livingston Ranch, 19 mi N Presidio, 3 (ASVRC); 5 mi SE Bandera Mesa, 4 (MWSU). *Brewster Co.:* 4 mi W Alpine, 1 (SRSU); Bird Mines, Altuda, 1 (SRSU); BGWMA, 35 (3 DMNHT, 8 MSU, 24 TNHC); 4 mi NE BGWMA hdqs., 1 (DMNHT); 5.5 mi SE BGWMA hdqs., 1 (DMNHT); Oak Creek, BBNP, 4,000 ft., 3 (2 TCWC, 1 BBNHA); 3 mi NE Terlingua, 2 (SRSU); Big 38 Mine, 3 mi W Terlingua ghost town, 3 (MWSU); Chisos Mts., BBNP, 1 (LACM); Kibee Springs, BBNP, 5,700 ft., 2 (1 FMNH, 1 MVZ); 0.5 mi NW Mt. Emory, Chisos Mts., BBNP, 1 (MVZ); Emory Peak Cave, BBNP, 7,100–7,500 ft., 14 (10 TCWC, 4 UMMZ); W slope Mt. Emory, 6,500 ft., BBNP, 2 (MVZ); Sul Ross Camp House, near Boquillas, 1 (TCWC); SE slope Mariscal Mt., 2,800 ft., BBNP, 1 (MVZ); Mariscal Mines, BBNP, 1 (BBNHA). *Val Verde Co.:* Fisher's Fissure, 2 mi W Langtry, 2 (TTU); Langtry, 1 (USNM/FWS); 33 mi N, 6 mi E Del Rio, 1 (TTU); E Painted Cave, 1 (USNM/FWS); mouth of Pecos River, 3 (2 MWSU, 1 TCWC).

Additional records: *Collingsworth Co.:* 4 mi N, 13 mi E Lutie (Hollander et al., 1987); 3 mi N, 12 mi E Lutie (Hollander et al., 1987). *Cottle Co.:* no specific locality (Davis, 1974). *Kimble Co.:* no specific locality (Davis, 1974). *Presidio Co.:* Sierra Vieja (Blair and Miller, 1949). *Edwards Co.:* Devil's Sink Hole (Blair, 1952b).

References. 2, 3, 4, 6, 7, 10, 12, 13, 15, 21, 24, 26, 27, 29, 31, 33, 50, 67, 73, 76, 82, 84, 89, 92, 97, 98, 99, 105, 110, 142, 145, 158, 163, 167, 181, 188, 189, 198, 201, 202, 217, 245, 247, 255, 256, 257, 262, 265, 267, 279, 304, 306, 315, 326, 342, 354, 356, 362, 393, 400, 401, 402, 414.

Plecotus rafinesquii Lesson, 1827
Rafinesque's Big-eared Bat

Description. This is a medium-sized bat (forearm = 40–46 mm, weight = 7–13 g) with very large ears that measure over 2.5 cm in length. A large, distinctive facial gland is present on each side of the snout. Pelage coloration is gray above and nearly white below. The fur of the ventral pelage is conspicuously bicolor—the bases of individual hairs are black with the tips being much lighter. Long toe hairs extend beyond the tips of the claws. Average external measurements are: total length, 95 mm; tail, 45 mm; hind foot, 11 mm; ear, 32 mm.

P. rafinesquii is similar in appearance to Townsend's big-eared bat (*Plecotus townsendii*). However, the two species are geographically separated in Texas, and they may be distinguished as discussed in the account for *P. townsendii.*

Plecotus rafinesquii. Rafinesque's Big-eared Bat. (Merlin D. Tuttle, Bat Conservation International)

Distribution. Rafinesque's big-eared bat is distributed throughout the forested areas of the southeastern United States and reaches the westernmost boundary of its range in extreme eastern Texas. Here, this bat is found in small numbers at scattered localities in the forested Pineywoods region, where it has been captured in June, July, September, October, November, and December (Schmidly, 1983). Late winter records (January–March) are not available, and it is not known if the bats hibernate in East Texas or migrate during this season. *P. rafinesquii* remains throughout winter and hibernates in caves and mines in the northern part of its range (Jones, 1977).

Subspecies. Texas specimens are referrable to the subspecies *P. r. macrotis* Le Conte, 1831, according to the latest taxonomic revision of the species (Handley, 1959).

Life History. Rafinesque's big-eared bat inhabits hollow trees, crevices behind loose tree bark, culverts, and abandoned buildings (Jones, 1977; Davis, 1974). Jones and Suttkus (1975) found this species using abandoned ammunition storage bunkers in Louisiana and an abandoned house in Mississippi. In Texas, *P. rafinesquii* has been captured in cisterns, barns, and unoccupied buildings (Schmidly, 1983). It often roosts in more open and well-lit areas than do most species and is regularly found in association with the eastern pipistrelle, *Pipistrellus subflavus* (Jones, 1977). *P. rafinesquii* hibernates in caves and similar shelters in the northern part of its range, but the winter habits of this bat in the southern U.S. have not been well documented.

P. rafinesquii may roost in colonies containing up to 100 individuals (Jones,

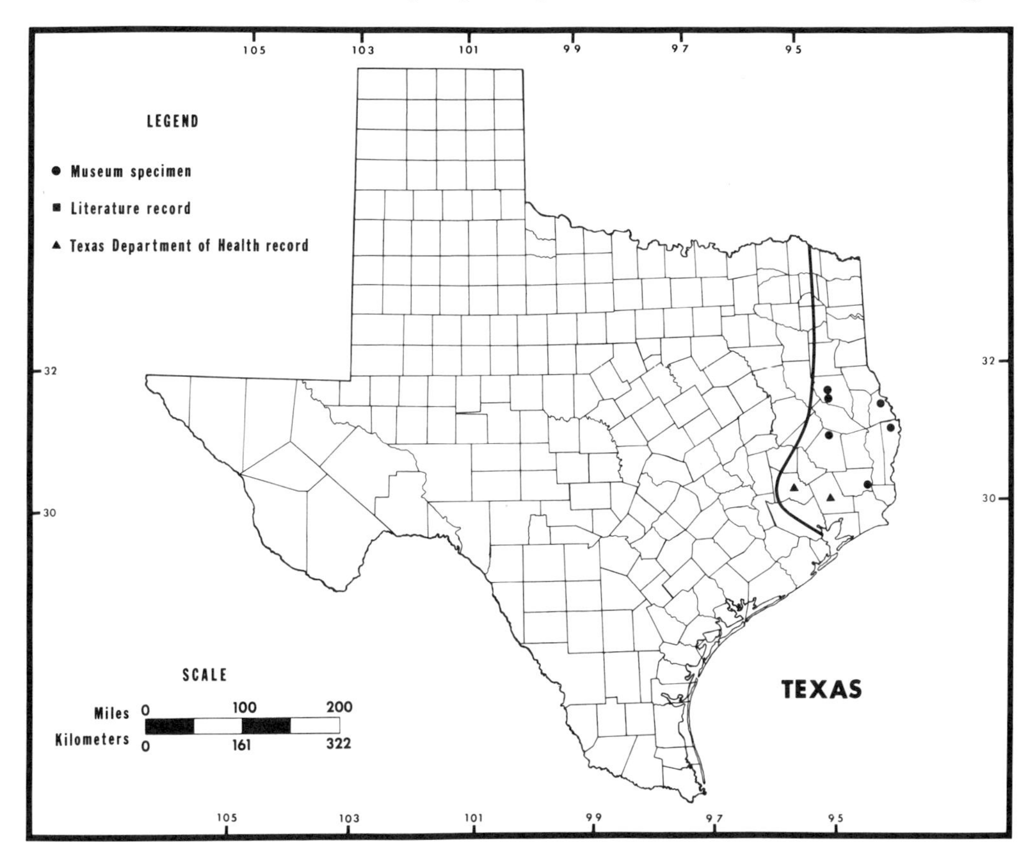

Map 22. Distribution of Rafinesque's big-eared bat, *Plecotus rafinesquii macrotis.*

1977), although single individuals are occasionally found. The sex ratios of colonies from East Texas are often unequal (Schmidly, 1983). Fifteen out of seventeen bats taken in a well in Nacogdoches County in October and November were males, whereas six out of seven bats captured in a barn in Polk County in November were females. Jones (1977) concluded that no specific differences in roost site selection were evident between the sexes, and frequent movement between roosts may account for the unequal sex ratios observed in East Texas colonies.

Rafinesque's big-eared bats emerge late in the evening to forage and, like other big-eared bats, they are very strong and agile fliers. Specific prey items are not known but small, night-flying insects, especially moths, probably provide the bulk of their diet.

Breeding probably occurs in fall and winter and a single young is produced after the nursery colonies form in spring. Parturition occurs from late May to early June in the northern part of the range and is slightly earlier in the south (Jones, 1977). The young are born naked and with the eyes closed; by three weeks of age the eyes have opened, permanent dentition is present, and the young bats are volant (Schmidly, 1983). Adult size is reached at approximately one month.

Only two specimens of *P. rafinesquii* have been reported to the Texas Department of Health and one of these tested positive for rabies.

Specimens examined (44). *Nacogdoches Co.:* Nacogdoches, 14 (SFASU); 10 mi S Nacogdoches, 2 (SFASU); 10 mi SE Nacogdoches, 10 (5 SFASU, 5 TCWC); 12 mi S Nacogdoches, 2 (SFASU); 15 mi SE Nacogdoches, 2 (SFASU). *Sabine Co.:* 3 mi NE Milam, 3 (SFASU). *Newton Co.:* 10.3 mi N Burkeville, 1 (TCWC); 10 mi N Burkeville, 1 (TCWC); 9.5 mi N Burkeville, 1 (TCWC); 7 mi N Burkeville, 1 (TCWC); 6 mi N Burkeville, 1 (TCWC). *Polk Co.:* 1 mi S Neches River, 3 mi E hwy. 59, 5 (3 SFASU, 2 TCWC). *Hardin Co.:* 7 mi SE Silsbee, 1 (TCWC).

References. 4, 7, 12, 13, 20, 21, 25, 26, 27, 29, 47, 67, 82, 217, 260, 265, 299, 315, 363, 364.

Antrozous pallidus (Le Conte, 1856)
Pallid Bat

Description. This is a medium-sized bat (forearm=48–60 mm, weight=12–17 g) with broad wings and large ears that are approximately 2.5 cm in length and slightly over 1.25 cm in width. The snout is blunt and lacks the two facial glands found in the genus *Plecotus*. Pelage coloration in *A. pallidus* is light yellow above, sometimes washed with brown or gray, and pale cream to white below. Females are often slightly larger than males. Average external measurements are: total length, 106 mm; tail, 44 mm; hind foot, 11 mm; ear, 28 mm.

The pallid bat is a distinctive bat, not easily confused with other species. Big-eared bats of the genus *Plecotus* (*P. townsendii* and *P. rafinesquii*) are distinguished by the presence of nose lumps and darker pelage, and the spotted bat (*Euderma maculatum*) is easily identified by its unique markings. Only *A. pallidus* has the combination of large ears and pale yellowish color.

Distribution. The pallid bat is known in Texas from the High Plains, Rolling Plains, Trans-Pecos, Edwards Plateau, and South Texas Plains regions. This is one of the most abundant bats of the Trans-Pecos, inhabiting mountainous areas, intermontane basins, and lowland desert scrub habitats at elevations ranging from 600 to 1,800 m. It is considerably less abundant toward the eastern margin of its range on the Edwards Plateau.

Little is known about the migratory habits of this species. Pallid bats have been collected extensively from late March through September in Texas, but they have not been obtained during the winter. *A. pallidus* is not known to perform long migrations and is thought to hibernate throughout much of its summer range. The paucity of winter records, however, may indicate otherwise.

Subspecies. In the past, all Texas specimens of *A. pallidus* have been assigned to the subspecies *A. p. pallidus* (Le Conte, 1856) (Martin and Schmidly, 1982). However, recent workers (Manning et al., 1988) have determined that pallid bats from the vicinity of the Red River and in the Panhandle of north-central Texas are referrable to another subspecies, *A. p. bunkeri* Hibbard, 1934, which is a somewhat larger and darker race with a limited range that extends into southwestern Kansas and western Oklahoma. The subspecies *A. p. pallidus* occupies the remainder

Antrozous pallidus. Pallid Bat. (Merlin D. Tuttle, Bat Conservation International)

of the pallid bat's range in Texas. Intergradation between the two subspecies may occur in the western Canadian River Valley (Manning et al., 1988).

Life History. Pallid bats inhabit rocky outcrop areas where they commonly roost in crevices, caves, mine tunnels, beneath rock slabs, and in buildings. Orr (1954) noted that they may occasionally roost in hollow trees in California. Colonies are usually small and contain from twelve to one hundred individuals. The winter habits are poorly known; it is thought that these bats hibernate on their summer range, although hibernacula have not been discovered. Bats found at the same sites with *A. pallidus* include the big brown bat (*Eptesicus fuscus*), Townsend's big-eared bat (*Plecotus townsendii*), Brazilian free-tailed bat (*Tadarida brasiliensis*), western pipistrelle (*Pipistrellus hesperus*), Yuma myotis (*Myotis yumanensis*), fringed myotis (*Myotis thysanodes*), and California myotis (*Myotis californicus*). With the exception of *T. brasiliensis*, which is often closely associated with *A. pallidus* (Orr, 1954), these bats typically segregate from *A. pallidus* at the roosting site.

Pallid bats employ a foraging strategy somewhat unusual among bats. They generally feed on insects greater than 17 mm in size, including both night-flying and flightless ground insects. The prevalence of noctural flightless insects in their diet makes it apparent that they alight on the ground to capture such prey. In other words, to some extent they are terrestrial foragers. Reports describing direct observations of pallid bats actively searching for flightless prey state that the in-

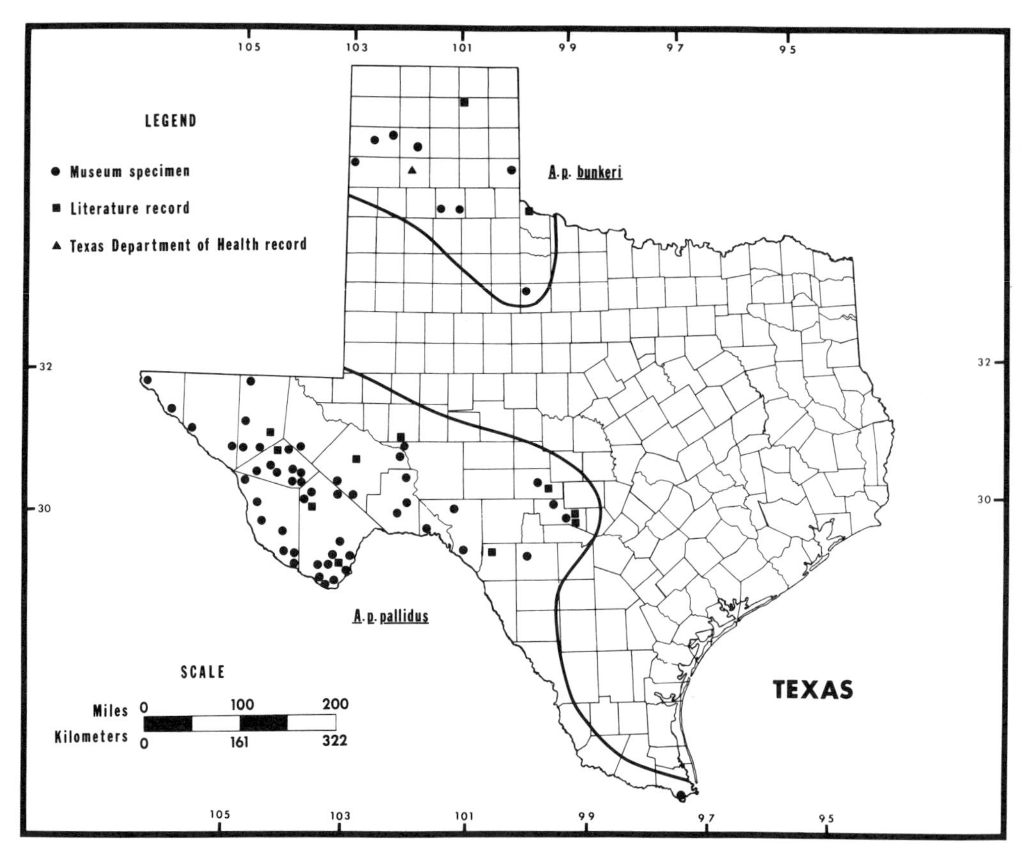

Map 23. Distribution of the two subspecies of the pallid bat, *Antrozous pallidus*.

dividual bat appears to fly at random over an area at one meter or less above the ground (Ross, 1967). It scouts, flying back and forth until it locates prey, presumably by sight. Then, abruptly it drops to the ground, searches briefly, grabs a victim in its mouth and takes off—not without some difficulty (Ross, 1967). Captured prey may be eaten on the ground or carried to a feeding station (Black, 1974).

Easterla (1973) examined the stomach contents of nine individuals from Big Bend National Park and found moths (22.2%), crickets (11.1%), froghoppers and leafhoppers (1.7%), antlions (8.9%), ground beetles (11.1%), and unidentified insects (45.0%). In a brief and general summary of their food habits, Orr (1954) listed specific prey in California as including Jerusalem crickets, June beetles, grasshoppers, ground beetles, and scorpions. Orr (1954) also noted that the average amount of food (mealworms) eaten by three captive pallid bats over a one-month period was 3.5 g per bat per night.

Ross (1961) detailed the specific prey consumed at four locations in Arizona and concluded that Jerusalem crickets were the most important component of their diet. Although pallid bats have often been portrayed as frequently attacking and eating scorpions, Ross (1961) found the remains of only one scorpion in his sample, leading him to speculate that scorpions are probably encountered by bats searching for Jerusalem crickets and flightless beetles. In a later report, Ross (1967) compiled a comprehensive food list for the pallid bat based on a survey of all published literature on the subject. His list included 54 different types of prey, of which large (20–70 mm), night-flying insects and flightless insects (20–50 mm), chiefly ground-dwelling arthropods, were the most prevalent.

In California, mating occurs in late October followed by parturition in late May to early June (Orr, 1954). Twente (1955a) recorded most births in late June in Kansas. In Texas, Manning et al. (1987) collected lactating females on May 19, June 19, and June 21 in Kimble County, and Easterla (1973) captured lactating females from June 11 to July 21 at Big Bend National Park. Parturition probably occurs from early May to mid-June in Texas.

Females may carry one to four embryos (Manning et al., 1987), but the birth of twins is usual. The length of gestation is fifty-three to seventy-one days (Orr, 1954). Newborn bats weigh about 3 g at birth and seem to develop more slowly than other species. The eyes open at eight to ten days, hair is evident at ten days, and the young are volant by six weeks of age (Orr, 1954). In Crockett County, Texas, Manning et al. (1987) captured two young *A. pallidus* whose stomachs were found to contain both insect remains and milk, suggesting that young pallid bats continue to nurse after they become volant.

Two pallid bats have been reported to the Texas Department of Health, and both tested negative for the rabies virus.

Specimens examined. *Antrozous pallidus bunkeri* (100). *Oldham Co.*: Griffin Ranch, 18 mi N, 1 mi W Adrian, 3 (TTU); 17 mi N, 1 mi W Adrian, 40 (TTU); Tascosa, 1 (USNM/FWS). *Potter Co.*: Fain Ranch, 16 mi N Amarillo, 39 (TTU). *Deaf Smith Co.*: 4.8 mi S, 4.9 mi E Glenrio, 4 (TTU). *Collingsworth Co.*: 3 mi N, 2 mi E Lutie, 1 (TTU). *Briscoe Co.*: 6.1 mi N, 0.1 mi W Quitaque, 1 (TTU); Caprock Canyons, 3 mi N Quitaque, 4 (TTU); Los Lingos Canyon, 2 (TTU). *Haskell Co.*: 8.5 mi SW Rochester, 1 (MWSU).

Additional records: *Roberts Co.*: 10 mi S, 15 mi E Spearman (Jones et al., 1988). *Oldham Co.*: 17–18 mi N, 1 mi W Adrian (Jones et al., 1988); 17–18 mi N, 1 mi E Adrian (Hollander et al., 1987); Tascosa (Bailey, 1905; Jones et al., 1988). *Potter Co.*: 16 mi N Amarillo (Hollander et al., 1987; Jones et al., 1988). *Collingsworth Co.*: 3 mi N, 12 mi E Lutie (Hollander et al., 1987). *Hardeman Co.*: 20 mi N Goodlett (Manning et al., 1987).

Antrozous pallidus pallidus (528). *El Paso Co.*: head of McKelligan Canyon, 4,700 ft., 3 (KU); Ft. Bliss, 2 (1 KU, 1 MSB); El Paso, 4 (2 FMNH, 2 USNM/FWS); 8 mi E Fabens, 4,000 ft., 4 (KU). *Hudspeth Co.*: Ft. Hancock, 13 (USNM/FWS); 2 mi S, 2 mi E Esperanza, 2 (UTEP); Wind Canyon, Eagle Mts., 9 (7 SRSU, 2 UTEP). *Culberson Co.*: McKittrick Canyon, GMNP, 13 (TCWC); Pratt Lodge, McKittrick Canyon, GMNP, 2 (TTU); South McKittrick Canyon, GMNP, 1 (TTU); Smith Springs, GMNP, 3 (TTU); The Bowl, GMNP, 4 (TTU); 4 mi E Pine Springs, GMNP, 2 (TCWC); 35 mi N Van Horn, 1 (TCWC); 25 mi N Van Horn, 1 (TCWC); Van Horn, 1 (USNM/FWS); 20 mi E Van Horn, 1 (TCWC). *Reeves Co.*: 3 mi WNW Toyahvale, 14 (MWSU); Toyahvale, 2 (UMMZ). *Pecos Co.*: 8 mi N, 1.5 mi W. Sheffield, 1 (KU); 8 mi N, 0.5 mi W Sheffield, 1 (TTU); 22.2 mi N, 1.8 mi W Marathon, 3,600 ft., 18 (TTU); 14.6 mi N, 19.2 mi E Marathon, 4,500 ft., 2 (TTU).

Jeff Davis Co.: 17 mi W Balmorhea, Cherry Canyon, 1 (TCWC); 8 mi W jct. hwys. 166 and 118, 2, (SRSU); 3 mi E jct. hwys. 166 and 118, 1 (TTU); Limpia Creek, 16 mi NE Ft. Davis, 4,000 ft., 3 (TCWC); Sawtooth Mt., Davis Mts., 5,700 ft., 3 (2 DMNHT, 1 TTU); mouth of Madera Canyon, 4,400 ft., 2 (TCWC); Madera Canyon, 14 mi NW Ft. Davis, 6,000 ft., 3 (TCWC); 14 mi NE Ft. Davis, old Whittenburg ranch, 5 (LACM); Reynold's Ranch, near Rockpile Park, Davis Mts., 4 (ASVRC); 10 mi N Ft. Davis, 2 (TTU); 8.5 mi N Ft. Davis, 8 (TTU); 6 mi NE Ft. Davis, 1 (TTU); 3.5 mi NE Ft. Davis, 2 (TTU); 1 mi N Ft. Davis, 1 (TTU); 8 mi S jct. hwys. 166 and 118, 2 (1 SRSU, 1 TTU); 5.5 mi N Mt. Livermore, 1 (TCWC); 5 mi E Mt. Livermore, 1 (UMMZ); Limpia Canyon, 3.5 mi NE Ft. Davis, 2 (TTU); Fort Davis, Hospital Canyon, 1 (UMMZ); 3 mi S Ft. Davis, 1 (TTU); Musquiz Canyon, 17.5 mi N Alpine, 1 (DMNHT); Valentine, 7 (TCWC); 3 mi E jct. hwys. 166 and 505, 1 (TTU); DMSP, 1 (TTU); 8 mi S DMSP, 1 (TTU); Fraiser Canyon, 3 (TTU); *Crockett Co.*: 5 mi N Iraan, Pecos River, 7 (CCSU). *Terrell Co.*: 15 mi SW Sheffield, 2 (MWSU); Cy Bonner Ranch, 29 mi N, 7 mi E Dryden, 3 (TCWC); 2 mi N Dryden, 6 (TCWC). *Kimble Co.*: Junction, Texas Tech campus, 1 (CCSU); 8 mi E Junction, 6 (MWSU). *Presidio Co.*: ZH Canyon, Sierra Vieja, 9 mi W Valentine, 5 (1 TCWC, 4 TTU); 8 mi NE Candelaria, 7 (TCWC); Chinati Mts., 3 (TTU); 9 mi SW Valentine, 1 (CCSU); Pinto Canyon, Chinati Mts., 45 mi SW Marfa, 18 (4 TCWC, 4 TTU, 9 SRSU, 1 UIMNH); Chinati Mts., 12 mi E Ruidosa, 5,000 ft., 1 (TCWC); Chinati Ranch, 23 mi NW Presidio, 4 (TCWC); Harper Ranch, 37 mi S Marfa, 4,000 ft., 1 (TCWC); 5 mi SE Bandera Mesa, 30 (29 MWSU, 1 NTSU); 15 mi E Redford, 1 (MWSU); 30 mi SSE Redford, 8 (MWSU); 3 mi W Lajitas, 1 (ASVRC); 2 mi W Lajitas, 3 (MWSU). *Brewster Co.*: 13.2 mi N, 2.6 mi E Marathon, 1 (TTU); 12.4 mi N, 2.6 mi E Marathon, 5,200 ft., 1 (TTU); 12.4 mi N, 5 mi E Marathon, 5,400 ft., 1 (TTU); 11.5 mi N, 2 mi W Marathon, 1 (TTU); 10 mi W Alpine, 3 (LACM); Alpine, 1 (USNM/FWS); Paisano, 2 (USNM/FWS); 38 mi S, 14 mi E Marathon, 17 (11 TTU, 6 UIMNH); BGWMA, 200 (138 DMNHT, 33 LACM, 13 MSU, 4 TCWC, 12 TTU); 4 mi E jct. Maravillas Creek, Rio Grande, 1 (DMNHT); 7 mi S BGWMA, 7 (DMNHT); N base Rosillos Mts., 2 (UMMZ); Grapevine Springs, BBNP, 3,000 ft., 2 (TCWC); Oak Spring, W Side Chisos Mts., BBNP, 1 (UMMZ); Oak Creek, BBNP, 4,000 ft., 3 (2 TCWC, 1 BBNHA); Bonham Ranch, Government Springs Ranch House, BBNP, 3,950 ft., 2 (AMNH); Kibee Spring, Chisos Mts., 5,700 ft., 1 (FMNH); Terlingua Creek, 4 mi E Terlingua, 2,200 ft., 4 (TCWC); Terlingua Creek, 3 mi W Study Butte, 1 (SRSU); Nail's Ranch, E side Burro Mesa, BBNP, 3,500 ft., 9 (1 LACM, 3 MVZ, 5 TCWC); 1.5 mi NW Boquillas, BBNP, 1 (UMMZ); Boquillas Ranger Station, BBNP, 5 (4 TTU, 1 BBNHA); Boquillas, 1 (USNM/FWS); Rio Grande Village, BBNP, 1 (BBNHA); mouth of Santa Elena Canyon, BBNP, 2,100 ft., 8 (TCWC); Mariscal Mine, 1 (BBNHA); Johnson's Ranch, Rio Grande, BBNP, 2,100 ft., 3 (TCWC); K-bar Ranch, BBNP, 1 (TCWC); BBNP, 5 (TTU); Big Bend of the Rio Grande, 2,000 ft., 1 (MVZ). *Val Verde Co.*: Juno, 14 (MWSU); 1 mi E Langtry, 4 (CCSU); Painted Cave, mouth Devil's River, 12 (USNM/FWS); Comstock, 21 (USNM/FWS). *Kerr Co.*: Camp Mystic, 24 mi NNW Kerrville, 14 (TCWC); 8 mi SW Kerrville on Turtle Creek, 1 (TCWC). *Uvalde Co.*: 25.0 mi NW Uvalde, Cal Newton Ranch, 2 (TCWC). *Cameron Co.*: 6 mi SE Brownsville, palm grove, 1 (TTU).

Additional records.: *Culberson Co.*: 18 mi NW Kent (Baker and Patton, 1967). *Upton Co.*: 3 mi S, 5 mi E McCamey (Manning et al., 1987). *Pecos Co.*: 12 mi E Ft. Stockton (Baker and Patton, 1967). *Jeff Davis Co.*: 13 mi S Kent (Walton and Siegel, 1966). *Crockett Co.*: 5 mi N, 4 mi W Iraan (Manning et al., 1987). *Kimble Co.*: Texas Tech Univ. Center at Junction (Manning et al., 1987); 1 mi S Texas Tech Univ. Center at Junction (Manning et al., 1987); 5 mi S Texas Tech Univ. Center at Junction (Manning et al., 1987); 17 mi SE Junction (Manning et al., 1987). *Brewster Co.*: 21 mi S Alpine (Walton and Siegel, 1966); Giant Dagger Yucca Flats (Easterla, 1968). *Kerr Co.*: Kerrville (Blair, 1952); 16 mi S Kerrville (Muliak, 1943). *Kinney Co.*: no specific locality (Davis, 1974).

References. 1, 2, 3, 4, 6, 7, 10, 12, 15, 21, 24, 29, 31, 33, 46, 69, 76, 84, 89, 92, 93, 97, 99, 102, 103, 105, 113, 147, 152, 155, 158, 163, 167, 181, 185, 188, 198, 201, 226, 233, 244, 247, 256, 257, 261, 262, 263, 265, 267, 268, 270, 279, 282, 286, 287, 288, 290, 292, 298, 304, 306, 313, 314, 318, 319, 322, 331, 342, 355, 356, 362, 393, 400, 401, 402, 403, 410, 414.

FAMILY MOLOSSIDAE

Nearly ninety species of the family Molossidae, or free-tailed bats, are found worldwide, primarily in tropical and subtropical regions of the Old World and Mexico south through South America. These bats are medium to large-sized, insectivorous, and characterized by having a tail which extends beyond the free edge of the uropatagium.

Molossids are swift, strong fliers and often fly great distances between roosting and feeding sites. Many species also make extensive migrations between winter and summer ranges. Because of their narrow wings, free-tailed bats have difficulty taking off from the ground and often roost high in buildings, cliffs, and caves. They require a free-fall for take off to enable them to achieve sufficient momentum to sustain level flight.

In North America free-tailed bats occur from Canada to Mexico, but they are most common in the southern and southwestern regions of the United States. Of the six species that occur in these areas, four are known from Texas.

Tadarida brasiliensis (I. Geof. St.-Hilaire, 1824)
Brazilian Free-tailed Bat

Description. This is a medium-sized bat (forearm=36–46 mm; weight=11–15 g) with broad ears that do not join at the midline of the head, vertical wrinkles on the lips along the muzzle, and dark brown to dark gray pelage. The individual hairs of the pelage are uniform in coloration, not bicolor. Long bristles are present on the feet and, as with all molossids, the terminal third of the tail projects beyond the free edge of the interfemoral membrane. Average external measurements are: total length, 93 mm; tail, 33 mm; hind foot, 8 mm; ear, 16 mm.

Tadarida brasiliensis. Brazilian Free-tailed Bat. (John L. Tveten)

Distribution. Brazilian free-tailed bats are the most common species of bat in Texas and are abundant statewide. Best known from Carlsbad Caverns National Park in New Mexico, these highly colonial bats actually reach their greatest concentrations in the Edwards Plateau region of Texas, where enormous summer populations occur in several caves. Davis et al. (1962) estimated population numbers for many of these caves, with the highest numbers occurring at Bracken, Goodrich, Rucker, and Frio caves, each containing from 10 to 20 million bats during the summer population peak periods. Additionally, smaller colonies are known from all parts of the state. Davis et al. (1962) estimate that between 95.8 and 103.8 million Brazilian free-tailed bats occupy Texas caves during the summer months.

Over most of Texas their presence is seasonal, although they are nonmigratory and year-round residents of the eastern part of the state. Populations from western and central Texas are highly migratory, arriving abruptly in early spring to raise young and then departing equally abruptly in late fall. Winter records are uncommon, except in warehouses or old buildings where they sometimes overwinter.

Subspecies. Two subspecies of Brazilian free-tailed bat are known from Texas, according to the latest taxonomic revision of the species in the state (Schmidly et al., 1977). *T. b. cynocephala* (Le Conte, 1831) is a nonmigratory resident of the eastern one-fourth of the state, and *T. b. mexicana* (Saussure, 1860) is a highly migratory subspecies found throughout the remainder of the state. Morphologically, these two subspecies are distinguished by differences in several skull characteristics (i.e. greatest length of skull, zygomatic breadth, and breadth of cranium), all of which are larger in *T. b. cynocephala.*

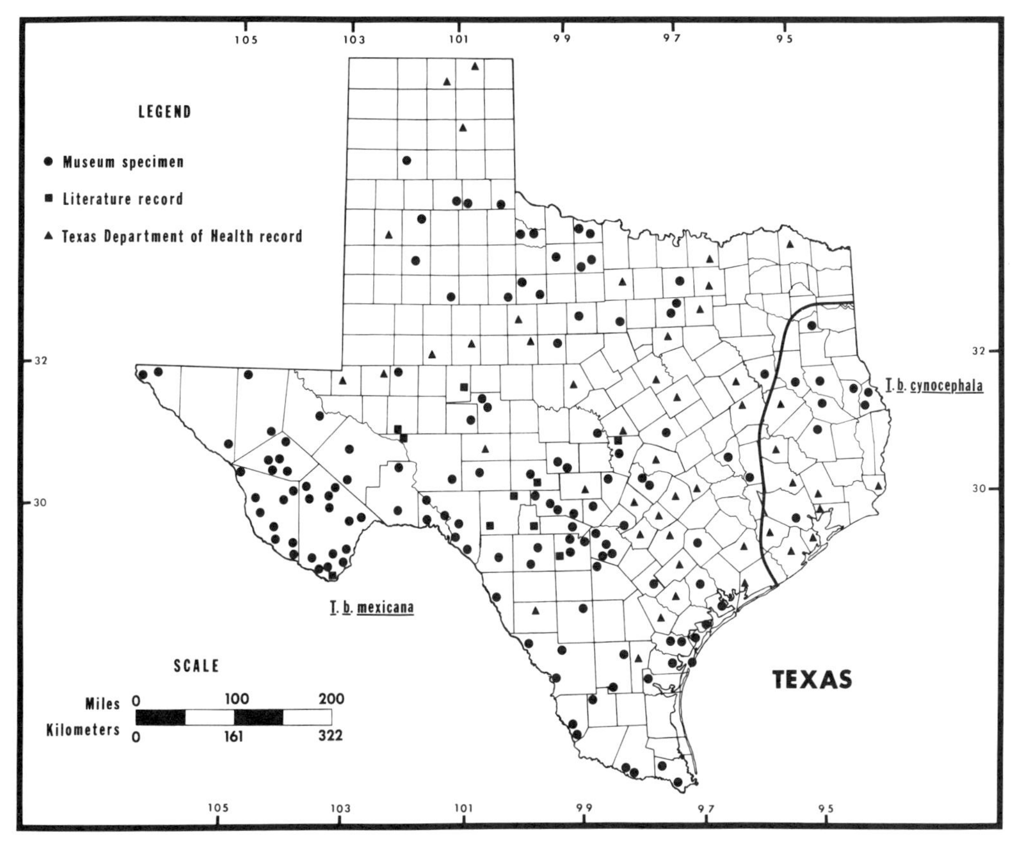

Map 24. Distribution of the two subspecies of the Brazilian free-tailed bat, *Tadarida brasiliensis.*

Most populations of the migratory subspecies, *mexicana,* have normally completed their move into Mexico prior to the onset of breeding, whereas *cynocephala* remains in the United States during the breeding season. This movement pattern would suggest that the two races are reproductively isolated over much of their range. However, overwintering populations of *mexicana* have been discovered in the area of contact between the two subspecies in southeastern Texas. Spenrath and LaVal (1974) discovered a colony of *mexicana* that overwintered at the old Animal Pavilion on the Texas A&M University campus in College Station, Brazos County, which is only 160 km from colonies of *cynocephala* in extreme eastern Texas. Schmidly et al. (1977) conducted a morphological analysis of cranial measurements from free-tailed bats near Navasota in Grimes County and found these specimens to be intermediate between *mexicana* and *cynocephala.* Thus, it appears that the two subspecies are not reproductively isolated and that they likely interbreed in this part of Texas.

Life History. This is among the best known of North American bats because of its widespread occurrence and abundance. In the west it is primarily a cave dweller, but in the east it predominantly occupies buildings. In fact, it is believed that few towns in the southeast lack a resident population of these bats.

Brazilian free-tailed bats also use rock fissures, crevices in structures such as bridges and signs, and cliff swallow nests during migration (Buchanan, 1958). Pitts and Scharninghausen (1986) discovered them using cliff swallow nests at Bandera in Medina County. The bats were found in association with the cave myotis (*Myotis velifer*) which actually used the swallow nests even more regularly than did *T. brasiliensis.* Roosts are almost always sufficiently high off the ground to allow an unobstructed drop of at least 3 m, enabling the animals to gain enough airspeed to sustain level flight.

Caves utilized by large colonies of these bats are often referred to as "guano caves," due to the tremendous amount of excrement that accumulates. Bat guano is an excellent fertilizer and bat caves have been mined for guano since the nineteenth century, continuing even to the present.

Populations of the subspecies *mexicana* are inefficient hibernators, which explains why they migrate to warmer latitudes in winter. Migratory movements of close to 1,300 km have been recorded (Villa-R. and Cockrum, 1962). Unlike their western relatives, populations of the subspecies *cynocephala,* which occur throughout the southeastern United States, are known to hibernate in human structures.

Carter (1962) summarized the features of buildings typically chosen by *cynocephala* for roost sites. Among these characteristics "(1) the building is not permanently occupied by man or there is an unoccupied space between the roost and the area occupied by man; (2) the buildings are not heated during the winter or there is an unheated area between the space used for hibernating and the heated area, such as an unoccupied second floor in a two-story building; (3) there is a fairly open area, such as an attic, which the bats usually occupy during the summer; and (4) there is an area of limited space, such as that between interior walls or between floors, that is protected from a moderately severe drop in outside temperature." He went on to state that churches and old two-story buildings provided admirable roosts for these bats.

In summer, Brazilian free-tailed bats choose the warmest areas of the roost site, an adaptation thought to contribute to quick development of the young. These bats are often found in association with other bat species. In the west these include the ghost-faced bat (*Mormoops megalophylla*), cave myotis (*Myotis velifer*), pallid bat (*Antrozous pallidus*), big brown bat (*Eptesicus fuscus*), and Yuma myotis (*Myotis yumanensis*). In the east, *T. b. cynocephala* is occasionally found in the same roost with the evening bat (*Nycticeius humeralis*) and the southeastern myotis (*Myotis austroriparius*). The different species almost always segregate and use different portions of the roost site.

Ross (1967) reported this small bat taking prey 2–10 mm in length and gave the following proportions of food items for eighty-eight specimens: moths, 34%; flying ants, 26.2%; June beetles and leaf beetles, 16.8%; leafhoppers, 15%; and true bugs, 6.4%. He further mentioned that *T. brasiliensis* feeds in groups of ten to thirteen individuals and often preys on densely swarming insects, a foraging method he called filter feeding. The huge summer colonies of these bats clearly would have a great impact on nearby insect populations. Davis et al. (1962) have estimated that these bats potentially could destroy from six thousand to eighteen thousand metric tons of insects annually in the state.

Males are sexually active from February to the first of April, just prior to the bats' arrival in Texas. Females ovulate in late March and produce a single young after a gestation period of between seventy-seven and eighty-six days. Occasionally, females may carry two embryos.

Parturition occurs within a much narrower time span than is exhibited by vespertilionid bats. Peak birthing occurs from early to mid-June, during which time 70% of the young bats are born within a span of ten days. More than 90% of the newborn bats have made their appearance within fifteen days. The baby bats weigh only about 2.5 g at birth and are blind, naked, and pink. Immediately after birth the baby bats are nursed and deposited on the ceiling of the roost. Nursery colonies may contain clusters of over a million baby bats. In the past it was thought that adult females made no attempt to locate their own young within this mass, but nursed the first two young encountered upon their nightly return to the roost (Davis et al., 1962). Recent studies, however, have shown that females do indeed recognize and nurse their own young (McCracken, 1984), a remarkable feat given the confusion within such huge swarms of bats.

Even in the warm nursery colonies, neonatals of *T. brasiliensis* mature more slowly than the young of most vespertilionid species. By six weeks of age, however, the young free-tails have attained adult-like dentition, pelage, and body weight and have attempted their first flights (Pagels and Jones, 1974). Females become sexually mature in their first year of life and may become pregnant as yearlings. Males, however, do not become sexually mature until eighteen to twenty-two months of age (Short, 1961a). As the young mature, the great nursery colonies begin to disperse and by late fall the bats head south for winter.

Known predators include great horned owls, barn owls, red-tailed hawks, Cooper's hawks, American kestrels, and Mississippi kites. Raccoons, opossums, and skunks will feed upon fallen bats, and some snakes, especially rat snakes, may climb cave walls to attack the bats.

With the exception of the red bat (*L. borealis*), the Brazilian free-tailed bat is reported more often to the Texas Department of Health (TDH) than any other bat. This species seems quite susceptible to the rabies virus. Of 430 bats reported to the TDH, 105 (24%) tested positive for rabies. This is the highest incidence of rabies known for any Texas bat—although the total number of confirmed rabies cases is miniscule in relation to the population of the bats as a whole.

Remarks. The taxonomic relationship of *T. b. brasiliensis* and *T. b. cynocephala* has been debated for years. The two bats are very similar morphologically and their primary differences seem to be ethological.

Based on morphological evidence, Schmidly et al. (1977) suggested that the two taxa were not reproductively isolated and accordingly should be regarded as subspecies of a single interbreeding species. However, in a recent biochemical genetics study among several populations of free-tailed bats in the southeastern United States, Robert Owen, a mammalogist at Texas Tech University, discovered fixed genetic markers between the two taxa. On this basis he believes the taxon *cynocephala* may represent a species distinct from *T. brasiliensis*.

Specimens examined. Tadarida brasiliensis cynocephala (325): *Gregg Co.*: Longview, 44 (5 LSUMZ, 39 TCWC). *Cherokee Co.*: Alto, 1 (SFASU). *Nacogdoches Co.*: Nacogdoches, 72 (59 SFASU, 13 TTU); Nacogdoches, SFA campus, 27 (20 SFASU, 7 TTU); 11 mi SW Nacogdoches, 1 (TTU). *San Augustine Co.*: 20 mi S Center, 1 (SFASU). *Sabine Co.*: 1 mi N Hemphill, 1 (SFASU); Pineland, 154 (TCWC). *Angelina Co.*: Lufkin, 2 (SFASU). *Polk Co.*: Corrigan, 1 (SFASU). *Harris Co.*: Houston, 21 (TCWC).

Additional records: *Harrison Co.*: Marshall (Cleveland et al., 1984).

Tadarida brasiliensis mexicana (1559): *Randall Co.*: Canyon, 1 (KU); *Briscoe Co.*: 6.1 mi N, 0.1 mi W Quitaque, 18 (TTU). *Hall Co.*: Red River, 5 mi N, 1 mi W Turkey, 1 (CM). *Childress Co.*: 21 mi S Childress, 5 (MWSU). *Hale Co.*: Plainview, 1 (TTU). *Foard Co.*: 16 mi NW Crowell, 2 (MWSU); 15 mi NW Crowell, 7 (MWSU); 3 mi N Crowell, 1 (MWSU); 20 mi W Crowell, 1 (MWSU). *Wichita Co.*: 5 mi N Electra, 1 (MWSU); 1 mi N Iowa Park, 1 (MWSU); Iowa Park, 3 (MWSU); Wichita Falls, 13 (MWSU). *Clay Co.*: Petriola, 1 (MWSU). *Lubbock Co.*: 1 mi N Lubbock, 1 (TTU); Lubbock, 2 (1 TTU, 1 UMMZ); 1 mi N, 0.75 mi W Wolfforth, 1 (TTU). *Baylor Co.*: Seymour, 2 (MWSU). *Archer Co.*: Archer City, 1 (MWSU); Lake Kickapoo, 1 (MWSU); Megargel, 1 (MWSU). *Denton Co.*: Denton, 1 (TCWC). *Garza Co.*: Justiceburg, 1 (TTU); 7 mi E Justiceburg, 22 (CCSU). *Stonewall Co.*: 9 mi S Aspermont, 3 (TTU). *Haskell Co.*: 10 mi W Rochester, 1 (MWSU); 6 mi E, 4 mi S, Haskell, 1 (MWSU). *Palo Pinto Co.*: Brazos, 20 (USNM/FWS). *Tarrant Co.*: 12 mi NW Grapevine, 2 (DMNHT); Fort Worth, 26 (FWMSH). *Callahan Co.*: Putnam, 1 (KU). *Midland Co.*: Midland, 1 (TTU). *Anderson Co.*: Palestine, 12 (TCWC). *El Paso Co.*: head McKelligan Canyon, 4,700 ft., 9 (KU); El Paso, 3 (1 TTU, 1 USNM/FWS, 1 UTEP); 8 mi E, 5 mi S City Hall, 3,700 ft., 1 (KU); Hueco Tanks, 1 (MSU). *Hudspeth Co.*: Eagle Mts., Wind Canyon, middle tank, 4 (UTEP). *Culberson Co.*: McKittrick Canyon, GMNP, 5,400 ft., 4 (TCWC); Smith Springs, GMNP, 1 (TTU); The Bowl, GMNP, 3 (TTU); 1 mi N Kent, 4,000 ft., 2 (TCWC). *Reeves Co.*: Pecos, 10 (7 ANSP, 3 USNM/FWS). *Tom Green Co.*: Carlsbad, 2 (TCWC); San Angelo, 1 (USNM/FWS). *San Saba Co.*: Gorman Falls Fish Camp, 3 (TCWC). *Pecos Co.*: Fort Stockton, 4 (USNM/FWS); 14.6 mi N, 19.2 mi E Marathon, 4,500 ft., 3 (TTU); 14.4 mi N, 19.3 mi E Marathon, 4,500 ft., 9 (CM). *Bell Co.*: Fort Hood Reservation, 2 mi W Kileen, 3 (CM). *Jeff Davis Co.*: Cherry Canyon, 17 mi W Balmorhea, 2 (TCWC); 8 mi W jct. hwys. 166 and 118, 2 (SRSU); 3 mi E jct. hwys. 166 and 118, 1 (TTU); Davis Mts., 3 (USNM/FWS); Old McGuire Homestead, Davis Mts. Resort, 2 (SRSU); DMSP, 1 (TTU); 3 mi E jct. hwys. 166 and 505, 1 (TTU); 2 mi NW Ft. Davis, 1 (UMMZ); 1 mi N Ft. Davis, 4 (UMMZ); 3.5 mi NE Ft. Davis, 1 (TTU); 3 mi S Ft. Davis, 1 (TTU). *Crockett Co.*: 25 mi SE Ozona, Bagget Ranch, 10 (TCWC). *Burnett Co.*: Burnett, 1 (USNM/FWS). *Brazos Co.*: Bryan, 1 (AMNH); College Station, 42 (1 AMNH, 1 KU, 20 TCWC, 20 TTU). *Mason Co.*: 15 mi SW Mason, Bat Cave, 2 (TCWC); James River Bat Cave, 8 (TTU). *Grimes Co.*: Navasota, 22 (TCWC). *Terrell Co.*: 15 mi SW Sheffield, 1 (MWSU); 6 mi S Dryden, 3 (MWSU). *Sutton Co.*: 4.1 mi N Sonora, 4 (TCWC); 3 mi NE

Sonora, 18 (TTU); 3 mi E Sonora, 1 (TTU). *Kimble Co.:* Junction, 3 (1 CCSU, 2 LACM); 8 mi E Junction, 1 (MWSU). *Presidio Co.:* 11 mi W Valentine, Sierra Vieja, 1 (TNHC); ZH Canyon, 9 mi W Valentine, Sierra Vieja, 3 (2 UIMNH, 1 TTU); 11 mi SW Valentine, Sierra Vieja, 1 (TNHC); 2 mi S Paisano, 1 (TCWC); 8 mi NE Candelaria, 54 (TCWC); 10 mi S Marfa, 4,300 ft., 1 (TCWC); San Esteban dam on Casa Piedra Rd., 1 (SRSU); Shely Ranch, Chinati Mts., 1 (SRSU); 45 mi SW Marfa, Pinto Canyon, Chinati Mts., 4 (TCWC); Chinati Mts., 5,000–7,000 ft., 9 (TCWC); Chinati Mts., 14 mi E Ruidosa, 2 (TTU); Livingston Ranch, Shafter Mine Area, 19 mi N Presidio, 3 (ASVRC); 6 mi E, 3 mi S Presidio, Rio Grande, 1 (KU); Bandera Mesa, 6 (MWSU); 1 mi E Bandera Mesa, 2 (MWSU); 30 mi SSE Redford 2 (MWSU); 1 mi E San Carlos, 11 (FMNH). *Brewster Co.:* 12.4 mi N, 5 mi E Marathon, 5,400 ft., 1 (TTU); Sul Ross campus, Alpine, 2 (SRSU); Alpine, 2 (USNM/FWS); Marathon, 1 (SRSU); Woodward Ranch Campground, 6 (SRSU); 16 mi S Marathon, 3,900 ft., 2 (TCWC); 30 mi S Longfellow, 1 (SRSU); 38 mi S, 14 mi E Marathon, 12 (UIMNH); BGWMA, 78 (1 TNHC, 4 TCWC, 62 DMNHT, 11 LACM); 4 mi E Maravillas Creek on Rio Grande, 12 (DMNHT); Cooper's Well, 47 mi S Marathon, 2,450 ft., 2 (MVZ); Government Spring, BBNP, 6 (AMNH); 2 mi E Terlingua, 1 (TTU); 3 mi E Terlingua, Terlingua Creek, 1 (SRSU); 4 mi E Terlingua, Terlingua Creek, 2,200 ft., 2 (TCWC); Study Butte, Rocky Run Creek, 2,200 ft., 4 (TCWC); Oak Spring, 4,000 ft., Chisos Mts., 1 (TCWC); Chisos Mts., upper Juniper Spring, BBNP, 1 (UMMZ); Basin, Chisos Mts., BBNP, 2 (1 SRSU, 1 BBNHA); Wilson's Tank, 5 mi SE Basin, 1 (FMNH); W base Chisos Mts., 3,500 ft., 6 (FMNH); Nail's Ranch, E side Burro Mesa, 3,500 ft., 6 (2 LACM, 3 MVZ, 1 TCWC); Boquillas, 1 (USNM/FWS); SE slope Mariscal Mt., Chisos Mts., 2,800 ft., 4 (MVZ); mouth Santa Elena Canyon, 2,100 ft., 1 (TCWC); Santa Elena Campground, 9 (TCWC); Big Bend of the Rio Grande, 3 (UMMZ); Stillwell Crossing, 2 (TTU). *Travis Co.:* 20.5 mi NW Austin, 1 (TNHC); Mansfield Dam, 2 (TNHC); Austin, 3 (1 TNHC, 2 USNM/FWS); Austin, UT Biology Bldg., 3 (TNHC). *Blanco Co.:* Davis Cave, Davis Ranch, 3 (TCWC). *Val Verde Co.:* Pecos R. near Pandale, 2 (TTU); 44 mi N, 6 mi W Del Rio, 2 (TTU); Presser's Ranch, 20 mi N Comstock, 138 (USNM/FWS); 2 mi W Langtry, Fisher's Fissure, 2 (TTU); Langtry, 2 (USNM/FWS); 0.75 mi E Langtry, Mile Canyon, 8 (TTU); 3 mi E Langtry, 2 (TTU); 10 mi W Comstock, 1,300 ft., 34 (TCWC); 12 mi W, 3 mi S Comstock, 24 (TTU); Shumla, 3 (LACM); mouth of Pecos, 1 (USNM/FWS); Del Rio, 15 (USNM/FWS). *Kerr Co.:* Ingram, 1 (USNM/FWS); Lacey's Ranch, Kerrville, 13 (USNM/FWS); Camp Verde, 14 (CCSU); 7 mi N Kerrville, 1 (TCWC); 2 mi W Kerrville, 7 (TCWC); 2 mi SW Kerrville, 1 (TCWC); 8 mi SW Kerrville, 12 (TCWC); 13 mi W Hunt, Kerr Wildlife Mgt. Area, 13 (TCWC), 5 mi W Hunt, 9 (TCWC); 8 mi SW Ingram, 4 (TCWC). *Kendall Co.:* 14 mi NE Comfort, 1 (TCWC). *Comal Co.:* 5 mi NW Bracken Cave, 29 (LSUMZ); 5 mi N Bracken Cave, 26 (LSUMZ); Bracken Cave, 10 mi SW New Braunfels, 133 (14 AMNH, 7 LACM, 75 TCWC, 37 TTU). *Bandera Co.:* Bandera, Ney Cave, 25 (TTU); Bat Cave, 7 mi SW Bandera, 1,710 ft., 5 (TCWC); 10 mi S Bandera, 1 (LACM). *Bexar Co.:* 2 mi NE Helotes, Scenic Loop Rd., 4 (TCWC); Helotes Creek, 4 (ANSP); San Antonio, 17 (13 AMNH, 1 FMNH, 2 USNM/FWS, 1 WMM); Fort Sam Houston Military Reservation, 6 (FMNH); Mitchell Lake, 10 mi S San Antonio, 5 (MVZ); Somerset 3 (KU). *Kinney Co.*: 7 mi E Brackettville, 1 (TNHC). *Uvalde Co.*: Concan, 17 (LACM); Frio Cave, 3 mi SE Concan, 125 (107 LACM, 18 TTU); 1 mi E, 4 mi S Concan, 31 (TCWC); 19 mi NW Sabinal, 1,200 ft., 10 (TCWC); 5 mi W Uvalde, Nueces River, 2 (TCWC); Uvalde, 10 (USNM/FWS). *Medina Co.*: Ney Cave, 23 (2 LACM, 21 TTU); 20 mi N Hondo, 27 (21 LACM, 6 USNM/FWS); Hondo, 1 (ANSP). *La Vaca Co.*: Hallettsville, 1 (TCWC). *Atascosa Co.*: 7 mi SW Somerset, 2 (TNHC). *Karnes Co.*: Runge, 2 (USNM/FWS). *Victoria Co.*: Victoria, 1 (USNM/FWS). *Maverick Co.*: Eagle Pass, 10 (1 AMNH, 9 USNM/FWS). *Calhoun Co.*: Indianola, 1 (USNM/FWS). *LaSalle Co.*: 8 mi NE Los Angeles, 1 (TCWC). *Aransas Co.*: Aransas Refuge, 1 (TCWC); Rockport, 1 (CCSU). *Webb Co.*: 6 mi S Encinal, 1 (TCWC); 45 mi NW Laredo, 2 (KU); Laredo, 4 (3 KU, 1 USNM/FWS), *San Patricio Co.*: Sinton, 24 (TCWC). *Duval Co.*: San Diego, 1 (ANSP); 7 mi E Hebbronville, 2 (CCSU). *Nueces Co.*: Corpus Christi, 5 (2 CCSU, 3 USNM/FWS). *Kleberg Co.*: Kingsville, 4 (2 TAIU, 2 TCWC); Padre Island National Seashore, Malaquite Visitor Center, 1 (CCSU). *Jim Hogg Co.*: 13.4 mi SSE Mirando City on hwy. 649, 1 (TCWC). *Zapata Co.*: 18 mi SSE Zapata, 307 ft., 7 (TCWC); Falcon Reservoir Dam, 5 (TCWC); *Starr Co.*: 15 mi NW Roma, 275 ft., 8 (TCWC); 13 mi NW Roma, Los Saenz, 290 ft., 1 (TCWC). *Hidalgo Co.*: 15 mi S Edinburg, 4 (TCWC); 5 mi S Mission, Anzalouas Dam, 1 (TCWC); Hidalgo, 1 (USNM/FWS); Bentsen State Park, 1 (LSUMZ); Santa Ana National Wildlife Refuge, 2 (USNM/FWS). *Cameron Co.*: Harlingen, 2 (MVZ); Brownsville, 10 (USNM/FWS).

Additional records: *Cottle Co.*: no specific locality (Davis, 1974). *Crosby Co.*: no specific locality (Davis, 1974). *Dickens Co.*: no specific locality (Davis, 1974). *Tarrant Co.*: 3 mi E Roanoke (Miller, 1948); Grapevine (Constantine, 1957). *Sterling Co.*: 3.5 mi S, 4 mi E Sterling City (Manning et al., 1987). *Upton Co.*: 3 mi S, 5 mi E McCamey (Manning et al., 1987). *Crockett Co.*: 5 mi S, 5 mi E McCamey (Manning et al., 1987). *Burnett Co.*: Beaver Creek Cave (Eads et al., 1957). *Llano Co.*: no specific locality (Davis, 1974). *Kimble Co.*: Texas Tech Univ. Center at Junction (Manning et al., 1987); 1 mi S, 0.3 mi W Junction (Manning et al., 1987); 5 mi S Texas Tech Univ. Center at Junction (Manning et al., 1987). *Presidio Co.*: Paisano (Constantine, 1957). *Brewster Co.*: 5 mi SE Chisos Mts., BBNP (Borell and Bryant, 1942); E Base Burro Mesa, BBNP, 3,500 ft. (Borell and Bryant, 1942); SE slope Mariscal Mt., BBNP, 2,800 ft. (Borell and Bryant, 1942). *Edwards Co.*: Devil's Sinkhole (Eads et al., 1957); Rucker Bat Cave, 35 mi SW Rocksprings (Selander and Baker, 1957). *Kerr Co.*: Hunt (Constantine, 1957). *Bexar Co.*: Camp Bullis (Brennan, 1945). *Real Co.*: Leaky (Manning et al., 1987). *Medina Co.*: Valdina Farms Sinkhole (Eads et al., 1957). *La Salle Co.*: Encinal (Constantine, 1957). *Refugio Co.*: no specific locality (Davis, 1974).

References. 1, 2, 3, 4, 6, 7, 10, 12, 13, 15, 20, 21, 24, 25, 26, 27, 29, 33, 62, 63, 66, 71, 73, 74, 75, 76, 77, 87, 92, 93, 96, 97, 98, 99, 102, 103, 105, 106, 107, 109, 110, 111, 113, 116, 118, 119, 120, 125, 126, 128, 129, 131, 135, 136, 138, 139, 142, 145, 146, 155, 157, 163, 167, 171, 172, 173, 181, 188, 197, 198, 201, 202, 203, 204, 205, 208, 225, 226, 227, 228, 229, 230, 231, 232, 234, 235, 236, 237, 238, 239, 240, 241, 242, 245, 248, 250, 251, 254, 255, 256, 257, 261, 262, 265, 267, 269, 272, 273, 274, 279, 288, 294, 295,

Nyctinomops femorosacca. Pocketed Free-tailed Bat. (Bruce J. Hayward)

296, 301, 304, 306, 307, 309, 312, 314, 324, 327, 328, 332, 336, 340, 342, 353, 355, 356, 362, 363, 365, 366, 368, 369, 370, 371, 372, 379, 380, 381, 382, 383, 386, 389, 391, 392, 393, 396, 407, 408, 411, 413, 414, 420.

Nyctinomops femorosacca (Merriam, 1884)
Pocketed Free-tailed Bat

Description. This is a medium-sized bat (forearm = 44–50 mm, weight = 10–14 g) with long, narrow wings, vertical wrinkles on the lips along the muzzle, and broad ears that are joined basally at the midline of the head. Pelage coloration is dark brown to gray above and below with the bases of the individual hairs nearly white. These bicolored hairs, and the joined ears, serve to distinguish this species from *T. brasiliensis,* which is similar in overall appearance but slightly smaller in size.

The common and scientific names of this bat refer to a shallow fold of skin on the underside of the uropatagium near the knee, which forms a pocket-like area. Average external measurements are: total length, 110 mm; tail, 37 mm; hind foot, 11 mm; ear, 24 mm.

Distribution. This rare bat was first recorded in Texas at Big Bend National Park by David Easterla in 1967 (Easterla, 1968). It typically occurs in arid, desert areas

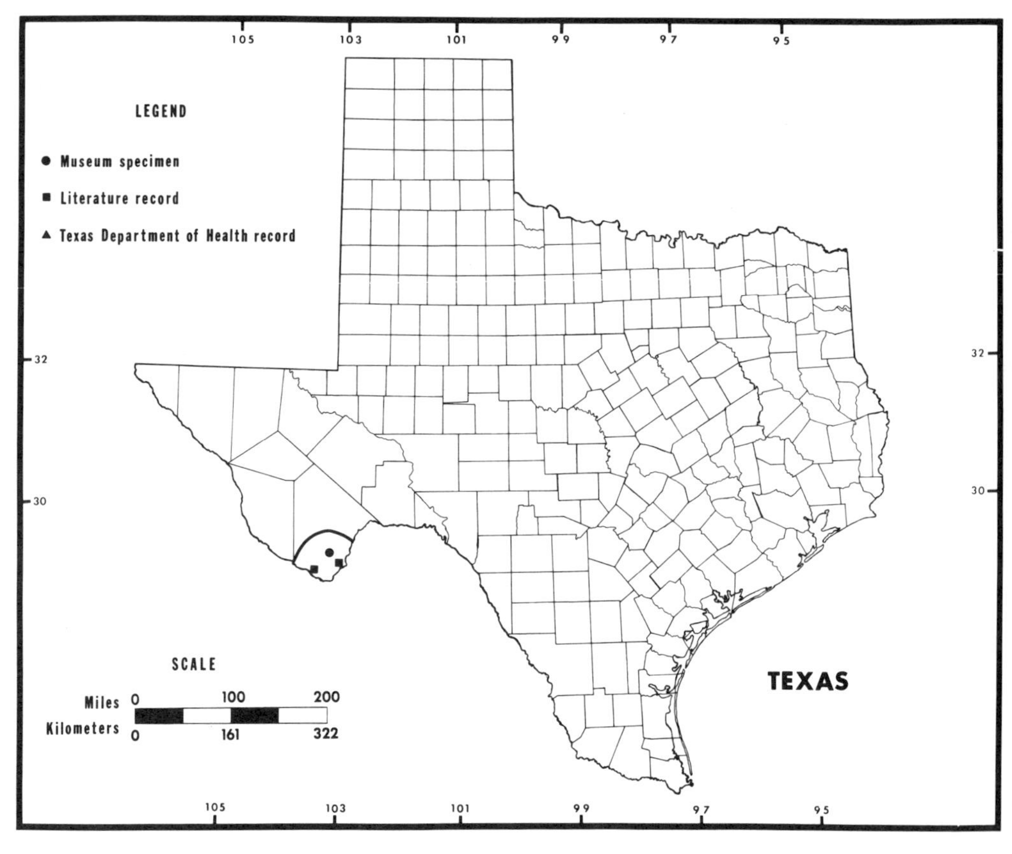

Map 25. Distribution of the pocketed free-tailed bat, *Nyctinomops femorosacca.*

with rugged canyons, rock outcrops, and high cliffs. Big Bend National Park represents the easternmost extension of its range in the United States. Nearest records to the park include Boquillas, Coahuila, Mexico; Alamos, Sonora, Mexico; and Carlsbad Caverns National Park, New Mexico. Studies conducted since its initial discovery at Big Bend have revealed that this bat is a well-established summer resident of the area from June through August.

Subspecies. *N. femorosacca* (Merriam, 1884) is monotypic and no subspecies are recognized.

Life History. Pocketed free-tailed bats are colonial roosters, but their colonies are small and generally contain fewer than one hundred individuals. They roost in caves and rock crevices of rugged canyon country, although they have also been reported in buildings. Nothing is known about their winter habits, and apparently they are absent from Texas during this period.

These bats typically leave their day roost late in the evening to forage, exhibiting swift, powerful flight. Easterla (1973) examined thirteen stomachs of individuals from Big Bend National Park and reported the following contents: moths

(36.9%), crickets (3.8%), flying ants (18.8%), stinkbugs (2.3%), froghoppers and leafhoppers (6.9%), lacewings (1.2%), and unidentified insects (30%).

A single young is born to the female in late June to early July. Easterla (1973) captured fifteen pregnant bats at Big Bend between June 10 and July 12 and found that each contained a single embryo. Lactating females were captured from July 7 to August 8.

Remarks. In her comprehensive study of the Molossidae, Freeman (1981) showed that the New World *Tadarida* (excluding *T. brasiliensis*) were morphologically distinct from Old World species of *Tadarida.* The generic name *Nyctinomops* has since been used for New World species formerly classified in the genus *Tadarida.*

Specimens examined (4). *Brewster Co.*: Giant Dagger Yucca Flats, BBNP, 3,000 ft., 1 (BBNHA); BBNP, no specific locality, 3 (1 BBNHA, 2 FMNH).

Additional records: Rio Grande Village, BBNP (Easterla, 1970); Castolon area, BBNP (Easterla, 1970).

References. 4, 7, 10, 12, 15, 21, 24, 33, 88, 130, 175, 177, 181, 185, 197, 198, 210, 245, 265, 356, 368, 411, 414.

Nyctinomops macrotis (Gray, 1839)
Big Free-tailed Bat

Description. This is the largest species (forearm = 58–64 mm) in the genus *Nyctinomops* with adults weighing from 22 to 30 g. It is characterized by large, broad ears that are joined basally at the midline of the head and extend beyond the tip of the snout when laid forward; vertical wrinkles on the lips along the muzzle; and reddish brown to dark brown or gray pelage coloration with somewhat glossy fur. Individual hairs are bicolored with the basal portions being white.

The large size of this bat serves to separate it from both *T. brasiliensis* and *N. femorosacca*, and the bicolored hair condition further distinguishes *N. macrotis* from *T. brasiliensis*. Average external measurements are: total length, 132 mm; tail, 53 mm; hind foot, 12 mm; ear, 28 mm.

Distribution. These bats have been recorded primarily from the Trans-Pecos region of Texas where they apparently are seasonal inhabitants of rugged, rocky country in both lowland and highland habitats. The only known nursery colony of this bat in the United States was discovered in the Chisos Mountains of Brewster County in Big Bend National Park by Borell (1939), who described the site as a horizontal crevice approximately 6 m long, 15 cm wide, and about 12 m above a talus slope. Borell estimated that approximately 130 bats occupied the colony in May and October. Subsequent investigators (Easterla, 1973) have been unable to relocate this colony.

Another nursery colony was suspected to occur in McKittrick Canyon of Guadalupe Mountains National Park, but it has yet to be located (LaVal, 1973).

Nyctinomops macrotis. Big Free-tailed Bat. (Merlin D. Tuttle, Bat Conservation International)

A day roost is known from Fern Canyon, Chihuahua, Mexico, which is very near Big Bend National Park (Easterla, 1972b). Big free-tailed bats have also been recorded from a small number of locations in the High Plains, the Gulf Prairies and Marshes, and the Post Oak Savannah.

With the exception of the single specimen from San Patricio County, which was found hanging on a screen door at the Welder Wildlife Refuge on December 23, 1959 (Raun 1961), there are no winter records of this species in Texas. In summer, segregation of sexes apparently occurs, as indicated by the fact that few males have been taken in the Trans-Pecos. David Easterla's (1973) study of bats in Big Bend National Park resulted in the capture of only one adult male out of 411 individuals.

Subspecies. *N. macrotis* (Gray, 1839) is monotypic and no subspecies are recognized.

Life History. This bat is uncommon and poorly known in Texas. It is believed to roost primarily in crevices of the rocky cliff country typical of western Texas, but it has also been reported in buildings. Winter habits of the species remain unknown, although they could possibly hibernate in the Trans-Pecos. Because they are such strong fliers and prone to wandering somewhat in the fall, these bats often turn up far from their normal range during the winter, as records from Sinton and Brazos counties and the High Plains attest.

Big free-tailed bats are powerful fliers, leaving their day roosts late in the evening to forage. Easterla (1973) examined the stomachs of forty-nine bats from Big Bend and found the contents to include moths (86.1%), crickets (6.7%), flying ants (4.1%), stinkbugs (1.3%), froghoppers and leafhoppers (0.1%), and unidentified insects (1.7%).

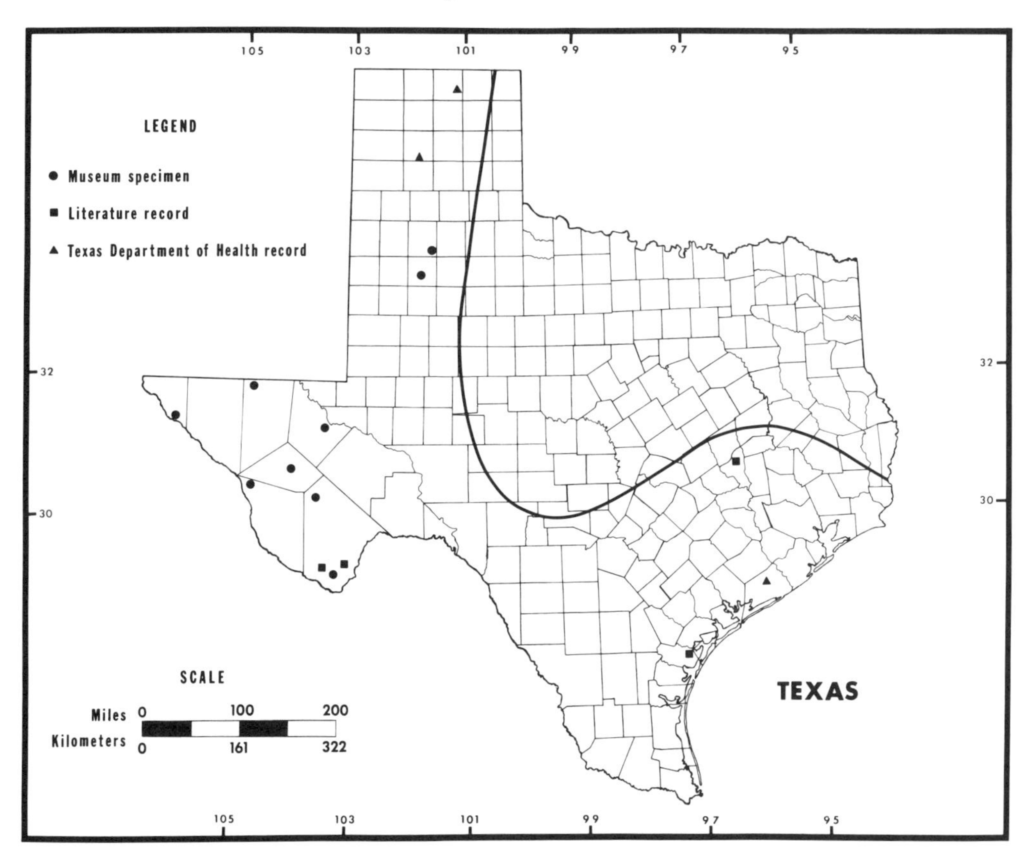

Map 26. Distribution of the big free-tailed bat, *Nyctinomops macrotis.*

Easterla (1973) captured thirty-six pregnant females in Big Bend between June 10 and July 7. Each contained a single embryo. Parturition is thought to occur from late June to early July, with a single young most common. Otherwise, the reproductive habits remain poorly known.

Two big free-tailed bats have been reported to the Texas Department of Health and both proved to be nonrabid.

Specimens examined (52). *Hale Co.: Petersburg,* 1 (TTU). *Lubbock Co.:* Lubbock, 1 (TTU). *El Paso Co.:* Tornillo, 1 (UTEP). *Culberson Co.:* McKittrick Canyon, GMNP, 5,600 ft., 13 (TCWC). *Reeves Co.:* Pecos, 1 (TCWC). *Jeff Davis Co.:* 0.5 mi SE Madera Canyon, 20 mi NW Ft. Davis, 5,900 ft., 2 (TTU). *Presidio Co.:* ZH Canyon, Sierra Vieja, 9 mi W Valentine, 2 (TTU). *Brewster Co.:* Alpine, 1 (TNHC); Pine Canyon, 5,600 6,000 ft., BBNP, 27 (4 FMNH, 13 MVZ, 2 KU, 8 TCWC); 0.25 mi above Boot Springs, BBNP, 1 (BBNHA); BBNP, 2 (1 KU, 1 BBNHA).

Additional records: *Brewster Co.:* Giant Dagger Yucca Flats, BBNP (Easterla, 1968); Terlingua Creek, opposite Study Butte (Constantine, 1961b). *Brazos Co.:* Bryan (Scarbrough, 1989). *San Patricio Co.:* Welder Wildlife Refuge Center (Raun, 1961).

References. 2, 4, 7, 10, 12, 15, 21, 24, 29, 31, 33, 70, 72, 104, 105, 113, 122, 134, 135, 167, 180, 181, 185, 197, 198, 201, 256, 257, 262, 265, 267, 279, 304, 310, 322, 338, 342, 356, 360, 362, 368, 411, 414, 415.

Eumops perotis (Schinz, 1821)
Western Mastiff Bat

Description. This is the largest (forearm, 72–82 mm) bat in the United States, with adults weighing around 65 g. Besides its large size, this species is characterized by large, broad ears that join basally at the midline of the head. The ears are not carried erect but slant forward and extend beyond the nose, nearly concealing the eyes. The lips are smooth along the muzzle and are not wrinkled vertically. Pelage coloration is dark gray dorsally and slightly paler below; individual hairs are bicolored, being nearly white at the base. Average external measurements are: total length, 171 mm; tail, 57 mm; hind foot, 17 mm; ear, 40 mm.

The large size of this bat and the absence of vertical wrinkles on the lips along the muzzle serve to distinguish this species from *Tadarida brasiliensis* and both species of *Nyctinomops* in Texas.

Distribution. This is a bat of the arid Southwest, where it inhabits rugged, rocky canyon country. In Texas, it has been recorded from localities close to the Rio Grande in Presidio, Brewster, and Val Verde counties. Although they have been collected only intermittently throughout the year in Texas, western mastiff bats are probably permanent residents of the Trans-Pecos.

Subspecies. Texas specimens are referrable to the subspecies *E. p. californicus* (Merriam, 1890), according to the most recent taxonomic review of the species (Eger, 1977).

Life History. Western mastiff bats have difficulty taking flight due to their large size and long, narrow wings. They live high in cliffs and must be able to free-fall before they can fly. Thus, roosting sites for this species always allow at least a three-meter unobstructed drop for initiating flight. Such sites include rock crevices and, occasionally, buildings. These bats roost in small colonies, commonly numbering only two or three to several dozen and generally less than a hundred individuals.

Ohlendorf (1972) observed a colony of mastiff bats in Capote Canyon which is located near Candelaria in Presidio County. This site is situated on the western face of the Sierra Vieja at an elevation of 1,130 m. The bats were roosting in a crevice formed by exfoliation of the nearly vertical rimrock. Openings were present on both the lower and upper sides of the slab and were unobstructed by vegetation. *Eumops* were observed at this site in January, February, May, June, July, August, September, and November. Ohlendorf estimated that at least seventy-one mastiff bats used this roost, and he stated the bats were usually quite vocal and audible at a distance of 180 m away from the roost. No other species were

Eumops perotis. Western Mastiff Bat. (Merlin D. Tuttle, Bat Conservation International)

seen around the *Eumops* roost, although *Pipistrellus hesperus, Eptesicus fuscus,* and *Tadarida brasiliensis* used other crevices within 90 m of the *Eumops* colony.

Ohlendorf's observations clearly suggest that *Eumops* is a year-round resident of the Trans-Pecos. Mastiff bats are active throughout winter, entering a daily torpor and then arousing at night to forage rather than entering the extended hibernation displayed by other bats.

These bats leave their day roosts late in the evening to forage. Easterla and Whitaker (1972) examined the stomachs of eighteen bats collected in Big Bend National Park and reported the contents to include moths (79.9%), crickets (16.5%), grasshoppers (2.8%), and unidentified insects (0.7%). Ross (1967) analyzed stomach contents of western mastiff bats from Arizona and found that they fed exclusively on the abdomens of large hawk moths up to 60 mm long. Bees, dragonflies, leafbugs, beetles, and cicadas have also been reported in their diet. These bats are not believed to use night roosts, but instead soar at great altitudes all night long so that they can feed over wide areas. Insects carried aloft by thermal currents probably furnish an important portion of their diet. The presence of flightless insects, such as crickets, in their diet is interesting, as these bats are unable to take off from the ground and, therefore, could not alight to capture such prey. These prey items could be picked from canyon walls as the bats forage.

Western mastiff bats breed in early spring with the length of gestation esti-

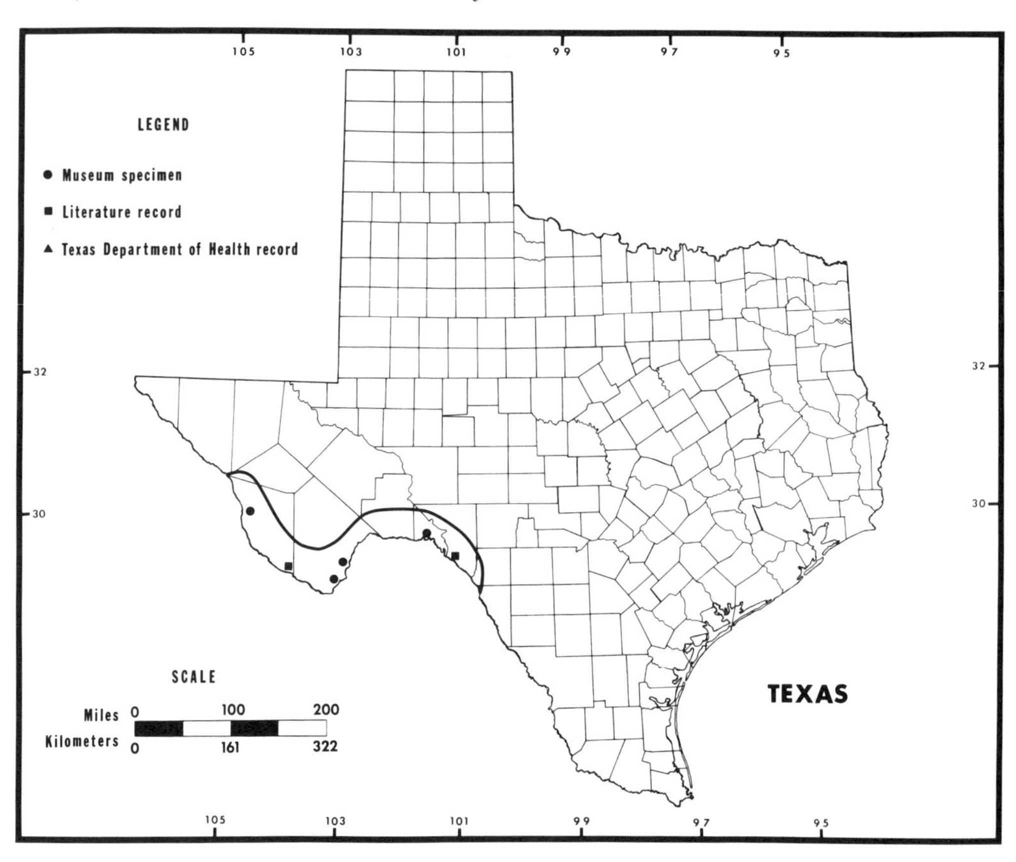

Map 27. Distribution of the western mastiff bat, *Eumops perotis*.

mated at eighty to ninety days. Parturition dates vary widely, however. Easterla (1973) captured twelve pregnant females between June 10 and 27 and one on July 21. Six of these carried a single embryo. Most births are believed to occur in June and July, with a single young generally the rule.

Adult male and female bats continue using the same roosts throughout the year, even during the period of parturition and development of the young. This is a rare situation among bats. Known predators of western mastiff bats include peregrin falcons, American kestrels, and red-tailed hawks.

Specimens examined (20). *Presidio Co.*: 8 mi NE Candelaria, 11 (TCWC); 3 mi NE Candelaria, 1 (TCWC). *Brewster Co.*: BGWMA, 3 (DMNHT); Chilicotal Mt., BBNP, 2 (TCWC); BBNP, no specific locality, 2 (BBNHA). *Val Verde Co.*: Langtry, 1 (USNM/FWS).

Additional records: *Presidio Co.*: Fresno Mine (Eads et al., 1957). *Brewster Co.*: Chilicotal Mt., BBNP (Constantine, 1961). *Val Verde Co.*: Pumpville (Barbour and Davis, 1969).

References. 3, 4, 7, 10, 12, 15, 21, 24, 29, 33, 76, 124, 135, 169, 181, 185, 186, 187, 197, 198, 265, 306, 317, 355, 356, 359, 406, 409, 411, 414.

4

Literature and References

The following is a comprehensive bibliography listing all of the general references, textbooks, reports, articles, and technical papers that were used in the preparation of this book. All references that are cited in the text have been included, as well as many that were not directly cited, but are nevertheless of importance to the study of Texas bats. The bibliography is divided into three sections: (1) general works, (2) *Mammalian Species*, and (3) technical papers.

General works include textbooks about bats, comprehensive guides to the mammalian fauna of geographical or political regions within and adjacent to Texas, general texts on mammalogy, and other bibliographies relevant to Texas bats. The *Mammalian Species* series consists of comprehensive summaries of knowledge about individual species of mammals. These accounts are published by the American Society of Mammalogists and contain the most current and complete information and references available on the respective species. All *Mammalian Species* accounts that are available for Texas bats have been included in this section and are highly recommended as a good starting point for further study of specific Texas bats. The section on technical papers is the most extensive and includes all documented records and studies of Texas bats as well as numerous references on bats from other parts of the United States and from Mexico that are of use in the study of Texas bats.

GENERAL WORKS

1. Alvarez, T. 1963. The Recent mammals of Tamaulipas, Mexico. *Univ. Kansas Publ. Mus. Nat. Hist.* 14:363–473.
2. Anderson, S. 1972. Mammals of Chihuahua: Taxonomy and distribution. *Bull. Am. Mus. Nat. Hist.* 148:151–410.
3. Baker, R. H. 1956. Mammals of Coahuila, Mexico. *Univ. Kansas Publ. Mus. Nat. Hist.* 9:125–335.
4. Barbour, R. W., and W. H. Davis. 1969. *Bats of America.* Lexington: Univ. Press of Kentucky. 286 pp.
5. Campbell, C. A. R. 1925. *Bats, mosquitoes, and dollars.* Boston: The Stratford Co. 262 pp.

6. Dalquest, W. W., and N. V. Horner. 1984. *Mammals of north-central Texas.* Wichita Falls: Midwestern State Univ. Press. 261 pp.
7. Davis, W. B. 1974. *The mammals of Texas.* Texas Parks and Wildlife Dept., Bull. no. 41. Austin. 294 pp.
8. Fenton, M. B. 1983. *Just bats.* Toronto: Univ. of Toronto Press. 165 pp.
9. Findley, J. S. 1987. *The natural history of New Mexican mammals.* Albuquerque: Univ. of New Mexico Press. 164 pp.
10. ———, A. H. Harris, D. E. Wilson, and C. Jones, 1975. *Mammals of New Mexico.* Albuquerque: Univ. of New Mexico Press. 360 pp.
11. Geluso, K. N., J. S. Altenbach, and R. C. Kerbo. 1987. *Bats of Carlsbad Caverns National Park.* Carlsbad: Carlsbad Caverns Natural History Association. 31 pp.
12. Hall, E. R. 1981. *The mammals of North America.* 2 vols. New York: John Wiley and Sons. 1,181 pp.
13. Harvey, M. J. 1986. *Arkansas bats: a valuable resource.* Arkansas Game and Fish Commission, Federal Aid Project E-1. 48 pp.
14. Hill, J. E., and J. D. Smith. 1984. *Bats, a natural history.* Austin: Univ. of Texas Press. 243 pp.
15. Hoffmeister, D. F. 1986. *Mammals of Arizona.* Tucson: The Univ. of Arizona Press. 602 pp.
16. Jones, J. K., Jr., D. C. Carter, and H. H. Genoways. 1979. Revised checklist of North American mammals north of Mexico, 1979. *Occas. Pap. Mus. Texas Tech Univ.* 62:1–17.
17. ———, and J. A. Homan. 1976. Contribution to a bibliography of Recent Texas mammals, 1961–1970. *Occas. Pap. Mus. Texas Tech Univ.* 41:1–21.
18. ———, C. J. Young, and D. J. Schmidly. 1985. Contribution to a bibliography of Recent Texas mammals, 1971–1980. *Occas. Pap. Mus. Texas Tech Univ.* 95:1–44.
19. Kunz, T. H., ed. 1988. *Ecological and behavioral methods for the study of bats.* Washington, D.C.: Smithsonian Inst. Press. 533 pp.
20. Lowery, G. H., Jr. 1974. *The mammals of Louisiana and its adjacent waters.* Baton Rouge: Louisiana State Univ. Press. 565 pp.
21. Mollhagen, T. 1970. A key to the bats of Texas and adjacent regions, with an annotated list. *Tex. Speleol. Surv.* 3:1–26.
22. Mumford, R. E., and J. O. Whitaker, Jr. 1982. *Mammals of Indiana.* Bloomington: Indiana Univ. Press. 537 pp.
23. Raun, G. G. 1962. A bibliography of the Recent mammals of Texas. *Tex. Mem. Mus. Bull.* 3:1–81.
24. Schmidly, D. J. 1977. *The mammals of Trans-Pecos Texas.* College Station: Texas A&M Univ. Press. 225 pp.
25. ———. 1983. *Texas mammals east of the Balcones Fault zone.* College Station: Texas A&M Univ. Press. 400 pp.
26. Schwartz, C. W., and E. R. Schwartz. 1981. *The wild mammals of Missouri.* Columbia: Univ. of Missouri Press. 356 pp.
27. Sealander, J. A. 1979. *A guide to Arkansas mammals.* Conway: River Road Press. 313 pp.

28. Slaughter, B. H., and D. W. Walton, eds. 1970. *About bats.* Dallas: Southern Methodist Univ. Press. 339 pp.
29. Taylor, W. P., and W. B. Davis. 1947. *The mammals of Texas.* Texas Game, Fish, and Oyster Commission, Bull. no. 27. Austin. 79 pp.
30. Tuttle, M. D. 1988. *America's neighborhood bats.* Austin: Univ. of Texas Press. 96 pp.
31. van Zyll de Jong, C. G. 1985. *Handbook of Candian mammals.* vol. 2. Ottawa: National Museums of Canada. 212 pp.
32. Vaughn, T. A. 1986. *Mammalogy, 3rd. ed.* Philadelphia: W. B. Saunders Co. 522 pp.
33. Villa-R., B. 1966. *Los murcielagos de Mexico.* Mexico: Universidad Nacional Autonoma de Mexico, Instituto de Biologia. 491 pp.
34. Wauer, R. H., and D. H. Riskind, eds. 1974. Symposium on the biological resources of the Chihuahuan Desert region, United States and Mexico. *Proc. Trans. Nat. Park Serv.* 3:1–658.
35. Wimsatt, W. A., ed. 1970. *Biology of bats.* New York: Academic Press. 406 pp.

MAMMALIAN SPECIES

36. Anderson, S. 1969. *Macrotus waterhousii. Mammalian Species* 1:1–4. Am. Soc. Mamm.
37. Arroyo-Cabrales, J., R. R. Hollander, and J. K. Jones, Jr. 1987. *Choeronycteris mexicana. Mammalian Species* 291:1–5. Am. Soc. Mamm.
38. Czaplewski, N. J. 1983. *Idionycteris phyllotis. Mammalian Species* 208:1–4. Am. Soc. Mamm.
39. Fenton, M. B., and R. M. R. Barclay, 1980. *Myotis lucifugus. Mammalian Species* 142:1–8. Am. Soc. Mamm.
40. Fitch, J. H., and K. A. Shump, Jr. 1979. *Myotis keenii. Mammalian Species* 121: 1–3. Am. Soc. Mamm.
41. ———, K. A. Shump, Jr., and A. U. Shump. 1981. *Myotis velifer. Mammalian Species* 149:1–5. Am. Soc. Mamm.
42. Fujita, M. S., and T. H. Kunz. 1984. *Pipistrellus subflavus. Mammalian Species* 228:1–6. Am. Soc. Mamm.
43. Greenhall, A. M., G. Joermann, U. Schmidt, and M. R. Seidel. 1983. *Desmodus rotundus. Mammalian Species* 202:1–6. Am. Soc. Mamm.
44. ———, U. Schmidt, and G. Joermann. 1984. *Diphylla ecaudata. Mammalian Species* 227:1–3. Am. Soc. Mamm.
45. Hensley, A. P., and K. T. Wilkins. 1988. *Leptonycteris nivalis. Mammalian Species* 307:1–4. Am. Soc. Mamm.
46. Hermanson, J. W., and T. J. O'Shea. 1983. *Antrozous pallidus. Mammalian Species* 213:1–8. Am. Soc. Mamm.
47. Jones, C. 1977. *Plecotus rafinesquii. Mammalian Species* 69:1–4. Am. Soc. Mamm.
48. ———, and R. W. Manning. 1989. *Myotis austroriparius. Mammalian Species* 332:1–3. Am. Soc. Mamm.

49. Kunz, T. H. 1982. *Lasionycteris noctivagans. Mammalian Species* 172:1–5. Am. Soc. Mamm.
50. ———, and R. A. Martin. 1982. *Plecotus townsendii. Mammalian Species* 175: 1–6. Am. Soc. Mamm.
51. Manning, R. W., and J. K. Jones, Jr. 1989. *Myotis evotis. Mammalian Species* 329:1–5. Am. Soc. Mamm.
52. O'Farrel, M. J., and E. H. Studier. 1980. *Myotis thysanodes. Mammalian Species* 137:1–5. Am. Soc. Mamm.
53. Shump, K. A., Jr., and A. U. Shump. 1982. *Lasiurus cinereus. Mammalian Species* 185:1–5. Am Soc. Mamm.
54. ———, and A. U. Shump. 1982. *Lasiurus borealis. Mammalian Species* 183:1–6. Am. Soc. Mamm.
55. Thomson, C. E. 1982. *Myotis sodalis. Mammalian Species* 163:1–5. Am. Soc. Mamm.
56. Warner, R. M. 1982. *Myotis auriculus. Mammalian Species* 191:1–3. Am. Soc. Mamm.
57. ———, and N. J. Czaplewski. 1984. *Myotis volans. Mammalian Species* 224: 1–4. Am. Soc. Mamm.
58. Watkins. L. C. 1972. *Nycticeius humeralis. Mammalian Species* 23:1–4. Am. Soc. Mamm.
59. ———. 1977. *Euderma maculatum. Mammalian Species* 77:1–4. Am. Soc. Mamm.
60. Webster, W. D., J. K. Jones, Jr., and R. J. Baker. 1980. *Lasiurus intermedius. Mammalian Species* 132:1–3. Am. Soc. Mamm.
61. Wilkins, K. T. 1987. *Lasiurus seminolus. Mammalian Species* 280:1–5. Am. Soc. Mamm.
62. ———. 1989. *Tadarida brasiliensis. Mammalian Species* 331:1–10. Am. Soc. Mamm.

TECHNICAL PAPERS

63. Adams, D. B., and G. M. Baer. 1966. Cesarian section and artificial feeding device for suckling bats. *J. Mamm.* 47:524–25.
64. Allan, P. F. 1947a. Bluejay attacks red bats. *J. Mamm.* 28:180.
65. ———. 1947b. Notes on Mississippi kites in Hemphill County, Texas. *Condor.* 49:88–89.
66. Allen, G. M. 1908. Notes on Chiroptera. *Bull. Harvard Mus. Comp. Zool.* 52:25–62.
67. ———. 1916. Bats of the genus *Corynorhinus. Bull. Harvard Mus. Comp. Zool.* 60:331–56.
68. ———. 1922. Bats from New Mexico and Arizona. *J. Mamm.* 3:156–62.
69. Allen, J. A. 1891. On a collection of mammals from southern Texas and northeastern Mexico. *Bull. Am. Mus. Nat. Hist.* 3:219–28.
70. ———. 1896a. Descriptions of ten new North American mammals. *Bull. Am. Mus. Nat. Hist.* 8:233–40.

71. ———. 1896b. On mammals collected in Bexar County and vicinity, Texas, by Mr. H. P. Attwater, with field notes by the collector. *Bull. Am. Mus. Nat. Hist.* 8:47–80.
72. Axtell, R. W. 1961. An additional record for the bat *Tadarida molossa* from Trans-Pecos Texas. *Southwestern Nat.* 6:52.
73. Baccus, J. T. 1971. The mammals of Baylor County, Texas. *Tex. J. Sci.* 22:177–85.
74. Baer, G. M., and G. M. Holquin. 1971. Breeding Mexican freetail bats in captivity. *Am. Midland Nat.* 85:515–17.
75. ———, and R. G. McLean. 1972. A new method of bleeding small and infant bats. *J. Mamm.* 53:231–32.
76. Bailey, V. 1905. *Biological survey of Texas.* N. Am. Fauna, vol. 25. Washington, D.C.: Dept. of Agriculture, Bureau of Biological Survey. 222 pp.
77. Baker, J. K. 1962. The manner and efficiency of raptor predation on bats. *Condor.* 64:500–504.
78. Baker, R. H. 1964. *Myotis lucifugus lucifugus* (Le Conte) and *Pipistrellus hesperus maximus* Hatfield in Knox County, new to north-central Texas. *Southwestern Nat.* 9:205.
79. ———, J. K. Jones, Jr., and D. C. Carter, eds. 1976. Biology of bats of the New World family Phyllostomatidae. Part III. *Spec. Publ. Mus. Texas Tech Univ.* 441 pp.
80. Baker, R. J., J. C. Patton, H. H. Genoways, and J. W. Bickham. 1988. Genic studies of *Lasiurus* (Chiroptera: vespertilionidae). *Occas. Pap. Mus. Texas Tech Univ.* 117:1–15.
81. Baker, R. J., and E. L. Cockrum. 1966. Geographic and ecological range of the long-nosed bats, *Leptonycteris*. *J. Mamm.* 47:329–31.
82. ———, and J. T. Mascarello. 1969. Chromosomes of some vespertilionid bats of the genus *Lasiurus* and *Plecotus*. *Southwestern Nat.* 14:249–51.
83. ———, T. Mollhagen, and G. Lopez. 1971. Notes on *Lasiurus ega*. *J. Mamm.* 52:849–52.
84. ———, and J. L. Patton. 1967. Karyotypes and karyotypic variation of North American vespertilionid bats. *J. Mamm.* 48:270–86.
85. Baker, R. K. 1954. A new bat (genus *Pipistrellus*) from northeastern Mexico. *Univ. Kansas Mus. Nat. Hist.* 7:583–86.
86. Barbour, R. W., and W. H. Davis. 1970. The status of *Myotis occultus*. *J. Mamm.* 51:150–51.
87. Baughman, J. L. 1951. The caves of Texas. *Tex. Game and Fish.* 9:2–7.
88. Benson, S. B. 1940. Notes on the pocketed free-tailed bat. *J. Mamm.* 21:26–29.
89. Bickham, J. W. 1979. Chromosomal variation and evolutionary relationships of vespertilionid bats. *J. Mamm.* 60:350–63.
90. ———. 1987. Chromosomal variation among seven species of lasiurine bats (Chiroptera: vespertilionidae). *J. Mamm.* 68:837–42.
91. Black, H. L. 1972. Differential exploitation of moths by the bats *Eptesicus fuscus* and *Lasiurus cinereus*. *J. Mamm.* 53:598–601.
92. ———. 1974. A north temperate bat community: structure and prey populations. *J. Mamm.* 55:138–57.

93. Blair, W. F. 1940. A contribution to the ecology and faunal relationships of the mammals of the Davis Mountain region, southwestern Texas. *Misc. Publ. Univ. Michigan Mus. Zool.*, no. 46. Ann Arbor. 39 pp.
94. ———. 1948. A color pattern aberration in *Pipistrellus subflavus subflavus*. *J. Mamm.* 29:178–79.
95. ———. 1950. The biotic provinces of Texas. *Tex. J. Sci.* 2:93–117.
96. ———. 1952a. Mammals of the Tamaulipan biotic province in Texas. *Tex. J. Sci.* 4:230–50.
97. ———. 1952b. Bats of the Edwards Plateau in central Texas. *Tex. J. Sci.* 4: 95–98.
98. ———. 1954. Mammals of the mesquite plains biotic district in Texas and Oklahoma, and speciation in the central grasslands. *Tex. J. Sci.* 6:235–64.
99. ———, and C. E. Miller, Jr. 1949. The mammals of the Sierra Vieja region, southwestern Texas, with remarks on the biogeographic position of the region. *Tex. J. Sci.* 1:67–92.
100. Bogan, M. A. 1974. Identification of *Myotis californicus* and *M. leibii* in southwestern North America. *Proc. Biol. Soc. Washington* 87:49–56.
101. ———. 1975. Geographic variation in *Myotis californicus* in the southwestern United States and Mexico. *U.S. Fish and Wildl. Serv. Res. Rpt.* 3:1–31.
102. ———, and D. F. Williams. 1970. Additional records of some Chihuahuan bats. *Southwestern Nat.* 15:131–34.
103. Borell, A. E. 1937. A new method of collecting bats. *J. Mamm.* 18:478–80.
104. ———. 1939. A colony of rare free-tailed bats. *J. Mamm.* 20:65–68.
105. ———, and M. D. Bryant. 1942. Mammals of the Big Bend area of Texas. *Univ. California Publ. Zool.* 48:1–62.
106. Brennan, J. M. 1945. Field investigations pertinent to Bullis fever. Preliminary report on the species of ticks and vertebrates occurring at Camp Bullis, Texas. *Tex. Rpts. Biol. and Med.* 3:112–21.
107. Buchanan, O. M. 1958. *Tadarida* and *Myotis* occupying cliff swallow nests. *J. Mamm.* 39:434–35.
108. Burnett, C. D. 1983. Geographic and secondary sexual variation in the morphology of *Eptesicus fuscus*. *Ann. Carnegie Mus. Nat. Hist.* 52:139–62.
109. Cagle, F. R. 1950. A Texas colony of bats, *Tadarida mexicana*. *J. Mamm.* 31: 400–402.
110. Caire, W., J. F. Smith, S. McGuire, and M. A. Royce. 1984. Early foraging behavior of insectivorous bats in western Oklahoma. *J. Mamm.* 65: 319–24.
111. Carter, D. C. 1962. The systematic status of the bat *Tadarida brasiliensis* (I. Geoffroy) and its related mainland forms. Ph.D. dissertation, Texas A&M Univ., College Station. 80 pp.
112. Chapman, B. R., and S. G. Spencer. 1987. Distributional records for six Texas mammals. *Tex. J. Sci.* 39:379–80.
113. Chapman, S. G. 1989. A survey of the distribution and ecological requirements of the bats of South Texas. M.S. thesis, Corpus Christi State Univ., Corpus Christi. 48 pp.
114. Choate, J. R., J. W. Dragoo, J. K. Jones, Jr., and J. A. Howard. 1986. Sub-

specific status of the big brown bat, *Eptesicus fuscus*, in Kansas. *Prairie Nat.* 18:43–51.

115. ———, and E. R. Hall. 1967. Two new species of bats, genus *Myotis*, from a Pleistocene deposit in Texas. *Am. Midland Nat.* 78:531–34.
116. Clark, D. R., Jr. 1981. Bats and environmental contaminants: a review. *U.S. Fish and Wildl. Serv. Spec. Sci. Rpt.* 235:1–27.
117. ———. 1988. How sensitive are bats to insecticides? *Wildl. Soc. Bull.* 16: 399–403.
118. ———, and J. C. Kroll. 1977. Effects of DDE on experimentally poisoned free-tailed bats (*Tadarida brasiliensis*): lethal brain concentrations. *J. Toxicology and Environ. Health* 3:893–901.
119. ———, C. O. Martin, and D. M. Swineford. 1975. Organochlorine insecticide residues in the free-tailed bat (*Tadarida brasiliensis*) at Bracken Cave, Texas. *J. Mamm.* 56:429–43.
120. Cleveland, A. G., J. T. Baccus, and E. G. Zimmerman. 1984. Distributional records and notes for nine species of mammals in eastern Texas. *Tex. J. Sci.* 35:323–26.
121. Cockerell, T. D. A. 1930. An apparently extinct Euglandina from Texas. *Proc. Colorado Mus. Nat. Hist.* 9:52–53.
122. Cockrum, E. L. 1952. The big free-tailed bat in Oklahoma. *J. Mamm.* 33:492.
123. ———. 1956. Homing, movements, and longevity of bats. *J. Mamm.* 37:48–57.
124. ———. 1960. Distribution, habitat and habits of the mastiff bat, *Eumops perotis*, in North America. *J. Arizona Acad. Sci.* 1:79–84.
125. ———. 1969. Migration in the guano bat, *Tadarida brasiliensis*. *Univ. Kansas Mus. Nat. Hist. Misc. Publ.* 51:303–36.
126. ———. 1970. Insecticides and guano bats. *Ecology*. 51:761–62.
127. Cokendolpher, J. C., D. L. Holub, and D. C. Parmley. 1979. Additional records of mammals from north-central Texas. *Southwestern Nat.* 24:376–77.
128. Constantine, D. G. 1948. Great bat colonies attract predators. *Nat. Speleological. Soc. Bull.* 10:100.
129. ———. 1957. Color variation and molt in *Tadarida brasiliensis* and *Myotis velifer*. *J. Mamm.* 38:461–66.
130. ———. 1958a. Remarks on external features of *Tadarida femorosacca*. *J. Mamm.* 39:437.
131. ———. 1958b. Bleaching of hair pigment in bats by the atmosphere in caves. *J. Mamm.* 39:513–20.
132. ———. 1958c. Ecological observations on lasiurine bats in Georgia. *J. Mamm.* 39:64–70.
133. ———. 1958d. Color variation and molt in *Mormoops megalophylla*. *J. Mamm.* 39:344–47.
134. ———. 1961a. Spotted bat and big free-tailed bat in northern New Mexico. *Southwestern Nat.* 6:92–97.
135. ———. 1961b. Locality records and notes on western bats. *J. Mamm.* 42: 404–405.
136. ———. 1962. Rabies transmission by nonbite route. *Public Health Rpts.* 77: 287–89.

137. ———. 1967a. Bat rabies in the southwestern United States. *Public Health Rpts.* 82:867–88.
138. ———. 1967b. Rabies transmission by air in bat caves. *Public Health Serv. Publ.* 1617:1–51.
139. ———. 1967c. Activity patterns of the Mexican free-tailed bat. *Univ. New Mexico Publ. Biol.* 7:1–79.
140. Creel, G. C. 1963. Bat as a food item of *Rana pipiens*. *Tex. J. Sci.* 15:104–106.
141. Cutter, W. L. 1959. The hoary bat in the panhandle of Texas. *J. Mamm.* 40: 442.
142. Czaplewski, N. J., J. P. Farney, J. K. Jones, Jr., and J. D. Druecker. 1979. Synopsis of bats of Nebraska. *Occas. Pap. Mus. Texas Tech Univ.* 61:1–24.
143. Dalquest, W. W. 1955. Natural history of the vampire bats of eastern Mexico. *Am. Midland Nat.* 53:79–87.
144. ———. 1967. Mammals of the Pleistocene Slaton local fauna of Texas. *Southwestern Nat.* 12:1–30.
145. ———. 1968. Mammals of north-central Texas. *Southwestern Nat.* 13:13–22.
146. ———. 1975. Vertebrate fossils from the Blanco local fauna of Texas. *Occas. Pap. Mus. Texas Tech Univ.* 30:1–52.
147. ———. 1978. Early Blancan mammals of the Beck Ranch local fauna of Texas. *J. Mamm.* 59:269–98.
148. ———. 1983. Mammals of the Coffee Ranch local fauna Hemphillian of Texas. *Pierce-Sellards Series* 38:1–41.
149. ———, and D. B. Patrick. 1989. Small mammals from the early and medial Hemphillian of Texas, with descriptions of a new bat and gopher. *J. Vert. Paleol.* 9:78–88.
150. ———, and E. R. Hall. 1947. Geographic range of the hairy-legged vampire in eastern Mexico. *Trans. Kansas Acad. Sci.* 50:315–17.
151. ———, E. Roth, and F. Judd. 1969. The mammal fauna of Schulze Cave, Edwards County, Texas. *Bull. Florida State Mus.* 13:205–76.
152. ———, and E. L. Roth. 1970. The pallid bat (*Antrozous*) of the Edwards Plateau. *Southwestern Nat.* 15:395–96.
153. ———, and F. B. Stangl, Jr. 1984a. The taxonomic status of *Myotis magnamolaris* Choate and Hall. *J. Mamm.* 65:485–86.
154. ———, and ———. 1984b. The Pleistocene mammals of Fowlkes Cave in southern Culberson County, Texas. Pp. 432–55 in Contributions in Quaternary vertebrate paleontology: a volume in memorial to John E. Guilday, ed. H. H. Genoways and M. R. Dawson. *Carnegie Mus. Nat. Hist. Spec. Publ.* 8:1–538.
155. ———, and ———. 1986. Post-Pleistocene mammals of the Apache Mountains, Culberson County, Texas, with comments on zoogeography of the Trans-Pecos Front Range. *Occas. Pap. Mus. Texas Tech Univ.* 104:1–35.
156. ———, and R. M. Carpenter. 1988. Early Pleistocene (Irvingtonian) mammals from the Seymour formation, Knox and Baylor counties, Texas, exclusive of Camelidae. *Occas. Pap. Mus. Texas Tech Univ.* 124:1–28.
157. Davis, R. B., C. F. Herreid II, and H. L. Short. 1962. Mexican free-tailed bats in Texas. *Ecol. Monographs* 32:311–46.

158. Davis, W. B. 1940. Mammals of the Guadalupe Mountains of western Texas. *Occas. Pap. Mus. Zool. Louisiana State Univ.* 7:69–84.
159. ———. 1944a. Status of *Myotis subulatus* in Texas. *J. Mamm.* 25:201.
160. ———. 1944b. Notes on Mexican mammals. *J. Mamm.* 25:370–402.
161. ———, and D. C. Carter. 1962a. Notes on Central American bats with description of a new subspecies of *Mormoops. Southwestern Nat.* 7:64–74.
162. ———, and D. C. Carter. 1962b. Review of the genus *Leptonycteris* (Mammalia: Chiroptera). *Proc. Biol. Soc. Washington* 75:193–98.
163. ———, and J. L. Robertson, Jr. 1944. The mammals of Culberson County, Texas. *J. Mamm.* 25:254–73.
164. Davis, W. H. 1959. Taxonomy of the eastern pipistrel. *J. Mamm.* 40:521–31.
165. ———. 1966. Population dynamics of the bat *Pipistrellus subflavus. J. Mamm.* 47:383–96.
166. Dooley, T. J. 1974. Bats of El Paso County, Texas, with notes on habitat, behavior, and ectoparasites. M.S. thesis, Univ. of Texas at El Paso. 63 pp.
167. ———, J. R. Bristol, and A. G. Canaris. 1976. Ectoparasites from bats in extreme west Texas and south-central New Mexico. *J. Mamm.* 57:189–91.
168. Dorsey, S. L. 1977. A reevaluation of two new species of fossil bats from Inner Space Caverns. *Tex. J. Sci.* 28:103–108.
169. Eads, R. B., J. E. Grimes, and A. Conklin. 1957. Additional Texas bat records. *J. Mamm.* 38:514.
170. ———, G. C. Menzies, and J. S. Wiseman. 1956. New locality records for Texas bats. *J. Mamm.* 37:440.
171. ———, J. S. Wiseman, J. E. Grimes, and G. C. Menzies. 1955. Wildlife rabies in Texas. *Public Health Rpts.* 70:995–1000.
172. ———, ———, and G. C. Menzies. 1955. Banding Mexican free-tailed bats. *J. Mamm.* 36:120–21.
173. ———, ———, and ———. 1957. Observations concerning the Mexican free-tailed bat, *Tadarida mexicana*, in Texas. *Tex. J. Sci.* 9:227–42.
174. Easterla, D. A. 1965. The spotted bat in Utah. *J. Mamm.* 46:665–68.
175. ———. 1968. First records of the pocketed free-tailed bat for Texas. *J. Mamm.* 49:515–16.
176. ———. 1970a. First records of the spotted bat in Texas and notes on its natural history. *Am. Midland Nat.* 83:306–308.
177. ———. 1970b. First record of the pocketed free-tailed bat for Coahuila, Mexico, and additional Texas records. *Tex. J. Sci.* 22:92–93.
178. ———. 1971. Notes on young and adults of the spotted bat, *Euderma maculatum. J. Mamm.* 52:475–76.
179. ———. 1972a. Status of *Leptonycteris nivalis* (Phyllostomatidae) in Big Bend National Park, Texas. *Southwestern Nat.* 17:287–92.
180. ———. 1972b. A diurnal colony of big freetail bats, *Tadarida macrotis* (Gray), in Chihuahua, Mexico. *Am. Midland Nat.* 88:468–70.
181. ———. 1973. Ecology of the 18 species of Chiroptera at Big Bend National Park, Texas. *Northwest Missouri State Univ. Studies* 34:1–165.
182. ———. 1975. The red bat in Big Bend National Park, Texas? *Southwestern Nat.* 20:418–19.

183. ———. 1976. Notes on the second and third newborn of the spotted bat, *Euderma maculatum*, and comments on the species in Texas. *Am. Midland Nat.* 96:499–501.
184. ———, and P. Easterla. 1969. America's rarest mammal. *National Wildlife* 7 (July): 15–18.
185. ———, and J. O. Whitaker. 1972. Food habits of some bats from Big Bend National Park, Texas. *J. Mamm.* 53:887–90.
186. Easterla, P., and D. A. Easterla. 1974. Rare glimpses of newborn bats. *Smithsonian.* 5 (May): 104–107.
187. Eger, J. L. 1977. Systematics of the Genus *Eumops* (Chiroptera: Molossidae). *Life Sciences Contrib. Royal Ontario Mus.* 110:1–69.
188. Farney, J., and E. D. Fleharty. 1969. Aspect ratio, loading, wingspan, and membrane areas of bats. *J. Mamm.* 50:362–67.
189. Fenton, M. B. 1969. The carrying of young by females of three species of bats. *Canadian J. Zool.* 47:158–59.
190. Findley, J. S. 1957. The hog-nosed bat in New Mexico. *J. Mamm.* 38:513–14.
191. ———. 1969. Biogeography of southwestern boreal and desert mammals. *Misc. Publ. Univ. Kansas Mus. Nat. Hist.* 51:113–28.
192. ———. 1972. Phenetic relationships among bats of the genus *Myotis. Syst. Zool.* 21:31–52.
193. ———, and C. Jones. 1964. Seasonal distribution of the hoary bat. *J. Mamm.* 45:461–70.
194. ———, and ———. 1967. Taxonomic relationships of bats of the species *Myotis fortidens, M. lucifugus,* and *M. occultus. J. Mamm.* 48:429–44.
195. ———, E. H. Studier, and D. E. Wilson. 1972. Morphologic properties of bat wings. *J. Mamm.* 53:429–44.
196. ———, and G. L. Traut. 1970. Geographic variation in *Pipistrellus hesperus. J. Mamm.* 51:741–65.
197. Freeman, P. W. 1981a. A multivariate study of the family Molossidae (Mammalia, Chiroptera): morphology, ecology, evolution. *Fieldiana Zool.* 7:1–173.
198. ———. 1981b. Correspondence of food habits and morphology in insectivorous bats. *J. Mamm.* 62:166–73.
199. Garner, H. W., and J. W. Bluntzer. 1975. Mammals of the Kansas-Texas boundary in Texas: distributional records of mammals along the boundary. *Tex. J. Sci.* 26:611–13.
200. Genoways, H. H., and R. J. Baker. 1988. *Lasiurus blossevillii* (Chiroptera: vespertilionidae) in Texas. *Tex. J. Sci.* 40:111–13.
201. ———, R. J. Baker, and J. E. Cornely. 1979. Mammals of the Guadalupe Mountains National Park, Texas. Pp. 271–332. In *Biological Investigations in the Guadalupe Mountains National Park, Texas,* ed. H. H. Genoways and R. J. Baker. *Proc. Trans. Nat. Park Serv.* 4:1–442.
202. George, J. E., and R. W. Strandtmann. 1960. New records of ectoparasites on bats in west Texas. *Southwestern Nat.* 5:228–29.
203. Glass, B. P. 1958. Returns of Mexican freetail bats banded in Oklahoma. *J. Mamm.* 39:435–37.

204. ———. 1959. Additional returns from free-tailed bats banded in Oklahoma. *J. Mamm.* 40:542–45.
205. ———. 1982. Seasonal movements of Mexican freetail bats *Tadarida brasiliensis mexicana* banded in the Great Plains. *Southwestern Nat.* 27:127–33.
206. ———, and R. J. Baker. 1968. The status of the name *Myotis subulatus* Say. *Proc. Biol. Soc. Washington* 81:257–60.
207. ———, and R. C. Morse. 1959. A new pipistrel from Oklahoma and Texas. *J. Mamm.* 40:531–34.
208. Goldman, E. A. 1926. Review of C. A. R. Campbell, *Bats, mosquitoes, and dollars. J. Mamm.* 7:136–38.
209. Gould, F. W., 1975. *Texas plants: A checklist and ecological summary.* College Station: Texas Agricultural Exp. Sta., Texas A&M Univ. 121 pp.
210. Gould, P. J. 1961. Emergence time of *Tadarida* in relation to light intensity. *J. Mamm.* 42:405–407.
211. Greenbaum, I. F., and C. J. Phillips. 1974. Comparative anatomy and general histology of tongues of long-nosed bats (*Leptonycteris sanborni* and *L. nivalis*) with reference to infestation of oral mites. *J. Mamm.* 55:489–504.
212. Greenhall, A. M. 1982. House bat management. *U.S. Fish and Wildl. Serv. Resour. Publ.* 143:1–33.
213. Hall, E. R., and W. W. Dalquest. 1950a. A synopsis of the American bats of the genus *Pipistrellus. Univ. Kansas Publ. Mus. Nat. Hist.* 1:591–602.
214. ———, and ———. 1950b. *Pipistrellus cinnamomeus* Miller 1902 referred to the genus *Myotis. Univ. Kansas Publ. Mus. Nat. Hist.* 1:584–90.
215. ———, and J. K. Jones, Jr. 1961. North American yellow bats, "Dasypterus," and a list of the named kinds of the genus *Lasiurus* Gray. *Univ. Kansas Publ. Mus. Nat. Hist.* 14:73–98.
216. Hamilton, W. J., Jr. 1933. The insect food of the big brown bat. *J. Mamm.* 14:155–56.
217. Handley, C. O., Jr. 1959. A revision of American bats of the genera *Euderma* and *Plecotus. Proc. U.S. Nat. Mus.* 110:95–246.
218. ———. 1960. Descriptions of new bats from Panama. *Proc. U.S. Nat. Mus.* 112:459–79.
219. Hargrave, L. L. 1944. A record of *Lasiurus borealis teliotis* from Arizona. *J. Mamm.* 25:414.
220. Harris, A. H. 1974. *Myotis yumanensis* in interior southwestern North America with comments on *Myotis lucifugus. J. Mamm.* 55:589–607.
221. Hatfield, D. M. 1936. A revision of the *Pipistrellus hesperus* group of bats. *J. Mamm.* 17:257–62.
222. Hayward, B. J. 1963. A maternity colony of *Myotis occultus. J. Mamm.* 44:279.
223. ———. 1970. The natural history of the cave bat *Myotis velifer. WRI-SCI (Western New Mexico Univ. Research in Science)* 1:1–74.
224. ———, and S. P. Cross. 1979. The natural history of *Pipistrellus hesperus* (Chiroptera: Vespertilionidae). *Office of Research, Western New Mexico Univ.* 3:1–36.

225. Henshaw, R. E. 1959. Responses of free-tailed bats to increases in cave temperature. *J. Mamm.* 41:396–98.
226. Hermann, J. A. 1950. The mammals of the Stockton Plateau of northeastern Terrell County, Texas. *Tex. J. Sci.* 3:368–93.
227. Herreid, C. F., II. 1958a. Four-thumbed free-tail bat. *J. Mamm.* 39:587.
228. ———. 1958b. Sexual dimorphism in teeth of the free-tailed bat. *J. Mamm.* 40:538–41.
229. ———. 1959a. Notes on a baby free-tailed bat. *J. Mamm.* 40:609–10.
230. ———. 1959b. Sexual dimorphism of the free-tailed bat. *J. Mamm.* 40:538–41.
231. ———. 1959c. Roadrunner a predator of bats. *Condor* 62:67.
232. ———. 1960. Comments on the odor of bats. *J. Mamm.* 41:396.
233. ———. 1961a. Notes on the pallid bat in Texas. *Southwestern Nat.* 6:13–20.
234. ———. 1961b. Snakes as predators of bats. *Herpetologica* 17:271–72.
235. ———. 1963a. Temperature regulation of Mexican free-tailed bats in cave habitats. *J. Mamm.* 44:560–73.
236. ———. 1963b. Metabolism of the Mexican free-tailed bat. *J. Cell. Comp. Physiol.* 61:201–207.
237. ———. 1963c. Temperature regulation and metabolism in Mexican freetail bats. *Science* 142:1573–74.
238. ———. 1967a. Temperature regulation, temperature preference and tolerance, and metabolism of young and adult free-tailed bats. *Physiological Zool.* 40:1–22.
239. ———. 1967b. Mortality statistics of young bats. *Ecology* 48:310–12.
240. ———, and R. B. Davis. 1960. Frequency and placement of white fur on free-tailed bats. *J. Mamm.* 41:117–19.
241. ———, and ———. 1966. Flight patterns of bats. *J. Mamm.* 47:78–86.
242. ———, ———, and H. L. Short. 1960. Injuries due to bat banding. *J. Mamm.* 41:398–400.
243. Hitchcock. H. B. 1965. Twenty-three years of bat banding in Ontario and Quebec. *Canadian Field Nat.* 79:4–14.
244. Hoffmeister, D. F. 1957. Review of the longnosed bats of the genus *Leptonycteris*. *J. Mamm.* 38:454–61.
245. ———. 1970. The seasonal distribution of bats in Arizona: a case for improving mammalian range maps. *Southwestern Nat.* 15:11–22.
246. Hollander, R. R., and J. K. Jones, Jr. 1987. A record of the western small-footed myotis, *Myotis ciliolabrum* Merriam, from the Texas Panhandle. *Tex. J. Sci.* 39:198.
247. ———, ———, R. W. Manning, and C. Jones. 1987. Noteworthy records of mammals from the Texas Panhandle. *Tex. J. Sci.* 39:97–102.
248. Humphrey, G. C., G. E. Kemp, and E. G. Wood. 1960. A fatal case of rabies in a woman bitten by an insectivorous bat. *Public Health Rpts.* 75:317–25.
249. Humphrey, S. R. 1975. Nursery roosts and community diversity of Nearctic bats. *J. Mamm.* 56:321–46.
250. Irons, J. V., R. B. Eads, J. E. Grimes, and A. Conklin. 1957. The public health importance of bats. *Tex. Rpts. Biol. and Med.* 15:292–98.

251. ———, ———, T. Sullivan, and J. E. Grimes. 1954. The current status of rabies in Texas. *Texas Rpts. Biol. and Med.* 12:489–99.
252. Izor, R. J. 1979. Winter range of the silver-haired bat. *J. Mamm.* 60:641–43.
253. Jackson, J. A., B. J. Schardien, C. D. Cooley, and B. E. Rowe. 1982. Cave myotis roosting in barn swallow nests. *Southwestern Nat.* 27:463–64.
254. James, P., and A. Hayse. 1962. Sparrow hawk preys on Mexican free-tailed bat at Falcon Reservoir. *J. Mamm.* 44:574.
255. Jameson, D. K. 1959. A survey of parasites of five species of bats. *Southwestern Nat.* 4:61–65.
256. Jones, C. 1965. Ecological distribution and activity periods of bats of the Mogollon Mountains area of New Mexico and adjacent Arizona. *Tulane Studies in Zool.* 12:93–100.
257. ———. 1966. Changes in populations of some western bats. *Am. Midland Nat.* 76:522–28.
258. ———, and J. Pagels. 1968. Notes on a population of *Pipistrellus subflavus* in southern Louisiana. *J. Mamm.* 49:134–39.
259. ———, and R. D. Suttkus. 1973. Colony structure and organization of *Pipistrellus subflavus* in southern Louisiana. *J. Mamm.* 54:962–68.
260. ———, and ———. 1975. Notes on the natural history of *Plecotus rafinesquii*. *Occas. Pap. Mus. Zool. Louisiana State Univ.* 47:1–14.
261. ———, ———, and M. A. Bogan. 1987. Notes on some mammals of North-Central Texas. *Occas. Pap. Mus. Texas Tech Univ.* 115:1–21.
262. Jones, J. K., Jr., E. D. Fleharty, and P. B. Dunnigan. 1967. The distributional status of bats in Kansas. *Univ. Kansas Publ. Mus. Nat. Hist.* 46:1–33.
263. ———, and H. H. Genoways. 1969. Holotypes of Recent mammals in the Museum of Natural History, The University of Kansas. *Univ. Kansas Misc. Publ. Mus. Nat. Hist.* 51:129–46.
264. ———, R. R. Hollander, and R. W. Manning. 1987. The fringed myotis, *Myotis thysanodes,* in west-central Texas. *Southwestern Nat.* 32:149.
265. ———, C. Jones, and D. J. Schmidly. 1988. Annotated checklist of Recent land mammals of Texas. *Occas. Pap. Mus. Texas Tech Univ.* 119:1–26.
266. ———, and M. R. Lee. 1962. Three species of mammals from western Texas. *Southwestern Nat.* 7:77–78.
267. ———, R. W. Manning, R. R. Hollander, and C. Jones. 1987. Annotated checklist of Recent mammals of Northwestern Texas. *Occas. Pap. Mus. Texas Tech Univ.* 111:1–14.
268. ———, ———, C. Jones, and R. R. Hollander. 1988. Mammals of the northern Texas Panhandle. *Occas. Pap. Mus. Texas Tech Univ.* 126:1–54.
269. Jones, R. S., and W. F. Hettler. 1959. Bat feeding by green sunfish. *Tex. J. Sci.* 11:48.
270. Judd, F. W. 1967. Notes on some mammals from Big Bend National Park. *Southwestern Nat.* 12:192–94.
271. ———, and D. J. Schmidly. 1969. Distributional notes for some mammals from western Texas and eastern New Mexico. *Tex. J. Sci.* 20:381–83.
272. Kohls, G. M., and W. L. Jellison. 1948. Ectoparasites and other arthropods occurring in Texas bat caves. *Nat. Speleol. Soc. Bull.* 10:116–17.

273. Krutzsch, P. H., and S. E. Sulkin. 1958. The laboratory care of the Mexican free-tailed bat. *J. Mamm.* 39:262–65.
274. ———, and A. H. Hughes. 1959. Hematological changes with torpor in the bat. *J. Mamm.* 40:547–54.
275. Kunath, C. E., and A. R. Smith, eds. 1968. The caves of the Stockton Plateau. *Tex. Speleol. Surv.* 3:1–111.
276. Kunz, T. H. 1973. Population studies of the cave bat (*Myotis velifer*): reproduction, growth, and development. *Occas. Pap. Mus. Univ. Kansas.* 15:1–43.
277. ———. 1974. Reproduction, growth, and mortality of the vespertilionid bat, *Eptesicus fuscus*, in Kansas. *J. Mamm.* 55:1–13.
278. LaVal, R. K. 1970. Infraspecific relationships of bats of the species *Myotis austroriparius*. *J. Mamm.* 51:542–52.
279. ———. 1973. Occurrence, ecological distribution, and relative abundance of bats in McKittrick Canyon, Culberson County, Texas. *Southwestern Nat.* 17:357–64.
280. ———, and M. L. LaVal. 1979. Notes on reproduction, behavior, and abundance of the red bat, *Lasiurus borealis*. *J. Mamm.* 60:209–12.
281. ———, and W. A. Shifflet. 1971. *Choeronycteris mexicana* from Texas. *Bat Research News* 12:40.
282. Le Conte, J. L. 1855. Observations on the American species of bats. *Proc. Acad. Nat. Sci. Philadelphia* 7:437.
283. Lee, T. E., Jr. 1987. Distributional record of *Lasiurus seminolus* (Chiroptera: vespertilionidae). *Tex. J. Sci.* 39:193.
284. Logan, L. E., and C. C. Black. 1979. The Quaternary vertebrate fauna of Upper Sloth Cave, Guadalupe Mountains National Park, Texas. Pp. 141–58. In *Biological investigations in the Guadalupe Mountains National Park, Texas*, ed. H. H. Genoways and R. J. Baker. *Trans. Nat. Park Serv.* 4:1–442.
285. Lundelius, E. L., Jr. 1967. Late-Pleistocene and Holocene faunal history of central Texas. Pp. 287–319. In *Pleistocene extinctions, the search for a cause*, ed. P. S. Martin and H. E. Wright, Jr. New Haven: Yale Univ. Press. 453 pp.
286. Manning, R. W., C. Jones, R. R. Hollander, and J. K. Jones, Jr. 1987. An unusual number of fetuses in the pallid bat. *Prairie Nat.* 19:261.
287. ———, ———, J. K. Jones, Jr., and R. R. Hollander. 1988. Subspecific status of the pallid bat, *Antrozous pallidus*, in the Texas Panhandle and adjacent areas. *Occas. Pap. Mus. Texas Tech Univ.* 118:1–5.
288. ———, J. K. Jones, Jr., R. R. Hollander, and C. Jones. 1987. Notes on distribution and natural history of some bats on the Edwards Plateau and in adjacent areas of Texas. *Tex. J. Sci.* 39:279–85.
289. ———, ———, and C. Jones. 1989. Comments on distribution and variation in the big brown bat, *Eptesicus fuscus*, in Texas. *Tex. J. Sci.* 41:95–101.
290. Martin, C. O. 1974. Systematics, ecology, and life history of *Antrozous* (Chiroptera: vespertilionidae). M.S. thesis, Texas A&M Univ., College Station, 256 pp.
291. ———. 1977. A noteworthy record of the silver-haired bat in southeast Texas. *Tex. J. Sci.* 28:356–57.

292. ———, and D. J. Schmidly. 1982. Taxonomic review of the pallid bat, *Antrozous pallidus* (Le Conte). *Spec. Publ. Mus. Texas Tech Univ.* 18:1–48.
293. Martin, R. A. 1972. Synopsis of late Pliocene and Pleistocene bats of North America and the Antilles. *Am. Midland Nat.* 87:326–35.
294. McCarley, W. H. 1959. The mammals of eastern Texas. *Tex. J. Sci.* 11:385–425.
295. ———, and W. N. Bradshaw. 1953. New locality records for some mammals of eastern Texas. *J. Mamm.* 34:515–16.
296. McCracken, G. F. 1984. Communal nursing in Mexican free-tailed bat maternity colonies. *Science* 223:1090–91.
297. Mearns, E. A. 1900. On the occurrence of a bat of the genus *Mormoops* in the United States. *Proc. Biol. Soc. Washington* 13:166.
298. Merriam, C. H. 1897. A new bat of the genus *Antrozous* from California. *Proc. Biol. Soc. Washington* 11:179–80.
299. Michael, E. D., and J. B. Birch. 1967. First Texas record of *Plecotus rafinesquii*. *J. Mamm.* 48:672.
300. ———, R. L. Wisennand, and G. Anderson. 1970. A Recent record of *Myotis austroriparius* from Texas. *J. Mamm.* 51:620.
301. Miller, F. W. 1948. The Mexican free-tailed bat in Tarrant County, Texas. *J. Mamm.* 29:418–19.
302. Miller, G. S., Jr. 1902. Note on the *Vespertilio incautus* of J. A. Allen. *Proc. Biol. Soc. Washington* 15:155.
303. ———, and G. M. Allen. 1928. The American bats of the genus *Myotis* and *Pizonyx*. *Bull. U.S. Nat. Mus.* 144:1–218.
304. Milstead, W. W., and D. W. Tinkle. 1959. Seasonal occurrence and abundance of bats (Chiroptera) in northwestern Texas. *Southwestern Nat.* 4:134–42.
305. Mitchell, R. W. 1970. Total number and density estimates of some species of cavernicoles inhabiting Fern Cave, Texas. *Ann. Speleol.* 25:73–90.
306. Mohr, C. E. 1948a. Texas cave bats. *Nat. Speleol. Soc. Bull.* 10:103–105.
307. ———. 1948b. Texas bat caves served in three wars. *Nat. Speleol. Soc. Bull.* 10:89–96.
308. ———. 1952. A survey of bat banding in North America, 1932–1951. *Nat. Speleol. Soc. Bull.* 14:3–13.
309. ———. 1972. The status of threatened species of cave dwelling bats. *Nat. Speleol. Soc. Bull.* 34:33–47.
310. Mollhagen, T. 1972. Distributional and taxonomic notes on some west Texas bats. *Southwestern Nat.* 17:427–30.
311. ———, and R. H. Baker. 1972. *Myotis volans interior* in Knox County, Texas. *Southwestern Nat.* 17:97.
312. Moore, W. 1949. Bat caves and bat bombs. *Turtox News* 26:262–65.
313. Morse, R. C., and B. P. Glass. 1960. The taxonomic status of *Antrozous bunkeri*. *J. Mamm.* 41:10–15.
314. Muliak, S. 1943. Notes on some bats of the southwest. *J. Mamm.* 24:269.
315. Nader, I. A., and D. F. Hoffmeister. 1983. Bacula of big-eared bats *Plecotus*, *Corynorhinus*, and *Idionycteris*. *J. Mamm.* 64:528–29.
316. O'Farrell, M. J., and E. H. Studier. 1973. Reproduction, growth, and develop-

ment in *Myotis thysanodes* and *M. lucifugus* (Chiroptera: vespertilionidae). *Ecology* 54:18–30.

317. Ohlendorf, H. M. 1972. Observations on a colony of *Eumops perotis* (Molossidae). *Southwestern Nat.* 17:297–300.

318. Orr, R. T. 1954. Natural history of the pallid bat, *Antrozous pallidus* (Le Conte). *Proc. California Acad. Sci.* 28:165–246.

319. ———, and G. S. Taboda. 1960. A new species of bat of the genus *Antrozous* from Cuba. *Proc. Biol. Soc. Washington.* 73:83–86.

320. Packard, R. L. 1966. *Myotis austroriparius* in Texas. *J. Mamm.* 47:128.

321. ———, and H. W. Garner. 1964. Records of some mammals from the Texas High Plains. *Tex. J. Sci.* 16:387–90.

322. ———, and F. W. Judd. 1968. Comments on some mammals from western Texas. *J. Mamm.* 49:535–38.

323. ———, and T. Mollhagen. 1971. Bats of Texas caves. Pp. 122–32. In *Natural history of Texas caves*, ed. E. L. Lundelius and B. H. Slaughter. Dallas: Gulf Natural History Assoc. 132 pp.

324. Pagels, J. F., and C. Jones. 1974. Growth and development of the free-tailed bat, *Tadarida brasiliensis cynocephala* (Le Conte). *Southwestern Nat.* 19:267–76.

325. Patton, T. H. 1963. Fossil vertebrates from Miller's Cave, Llano County, Texas. *Tex. Mem. Mus. Bull.* 7:1–41.

326. Pearson, O. P., M. R. Koford, and A. K. Pearson. 1952. Reproduction of the lump-nosed bat (*Corynorhinus rafinesquii*) in California. *J. Mamm.* 33: 273–320.

327. Peterson, R. L. 1946. Recent and Pleistocene mammalian fauna of Brazos County, Texas. *J. Mamm.* 27:162–69.

328. Petit, M. G. 1978. Imperiled bats of Eagle Creek Cave. *Nat. Hist.* 87 (March): 50–55.

329. Phillips, C. J. 1971. The dentition of glossophagine bats: development, morphological characteristics, variation, pathology, and evolution. *Misc. Publ. Univ. Kansas Mus. Nat. Hist.* 54:1–138.

330. ———, J. K. Jones, Jr., and F. J. Radovsky. 1969. Macronyssid mites in oral mucosa of long-nosed bats: occurrence and associated pathology. *Science.* 165:1368–69.

331. Pine, R. H., D. C. Carter, and R. K. LaVal. 1971. Status of *Bauerus* Van Gelder and its relationship to other nyctophiline bats. *J. Mamm.* 52:663–69.

332. Pitts, R. M., and J. J. Scharninghausen. 1986. Use of cliff swallow and barn swallow nests by the cave bat, *Myotis velifer*, and the free-tailed bat, *Tadarida brasiliensis*. *Tex. J. Sci.* 38:265–66.

333. Poche, R. M. 1975. New record of *Euderma maculatum* from Arizona. *J. Mamm.* 56:931–33.

334. ———, and G. L. Bailie. 1974. Notes on the spotted bat (*Euderma maculatum*) from southwest Utah. *Great Basin Nat.* 34:254–56.

335. ———, and G. A. Ruffner. 1975. Roosting behavior of male *Euderma maculatum* from Utah. *Great Basin Nat.* 35:121–22.

336. Randolph, N. M., and R. B. Eads. 1946. An ectoparasitic survey of mammals from La Vaca County, Texas. *Annals Entomological Soc. Am.* 39:597–601.
337. Raun, G. G. 1960. A mass die-off of the Mexican brown bat, *Myotis velifer*, in Texas. *Southwestern Nat.* 5:104–105.
338. ———. 1961. The big free-tailed bat in southern Texas. *J. Mamm.* 42:253.
339. ———. 1966. A heretofore unnoted collection of Texas mammals. *Tex. J. Sci.* 18:225–27.
340. ———, and J. K. Baker. 1959. Some observations of Texas cave bats. *Southwestern Nat.* 3:102–106.
341. Ray, C. E., and D. E. Wilson. 1979. Evidence for *Macrotus californicus* from Terlingua, Texas. *Occas. Pap. Mus. Texas Tech Univ.* 57:1–10.
342. Reddell, J. R. 1967. A checklist of the cave fauna of Texas. III—Vertebrata. *Tex. J. Sci.* 19:184–226.
343. ———. 1968. The hairy-legged vampire, *Diphylla ecaudata*, in Texas. *J. Mamm.* 49:769.
344. ———. 1971. A checklist of the cave fauna of Texas. VI—Additional records of Vertebrata. *Tex. J. Sci.* 22:139–58.
345. ———, ed. 1963. The caves of Val Verde County. *Tex. Speleol. Surv.* 1:1–53.
346. ———, ed. 1964. The caves of Comal County. *Tex. Speleol. Surv.* 2:1–60.
347. ———, ed. 1967. The caves of Medina County, Texas. *Tex. Speleol. Surv.* 3:1–58.
348. ———, and R. Finch, eds. 1967. The caves of Williamson County. *Tex. Speleol. Surv.* 2:1–61.
349. ———, and W. H. Russell, eds. 1963. The caves of Northwest Texas. *Tex. Speleol. Surv.* 1:1–56.
350. ———, and A. R. Smith, eds. 1965. The caves of Edwards County. *Tex. Speleol. Surv.* 2:1–70.
351. Rehn, J. A. 1902. A revision of the genus *Mormoops*. *Proc. Acad. Nat. Sci. Philadelphia.* 54:160–72.
352. Rice, D. W. 1957. Life history and ecology of *Myotis austroriparius* in Florida. *J. Mamm.* 38:15–32.
353. Ridlehuber, K. T., and N. J. Silvy. 1981. Texas rat snake feeds on Mexican freetail bat and wood duck eggs. *Southwestern Nat.* 26:70–71.
354. Rippy, C. L., and M. J. Harvey. 1965. Notes on *Plecotus townsendii virginianus* in Kentucky. *J. Mamm.* 46:499.
355. Ross, A. J. 1961. Notes on food habits of bats. *J. Mamm.* 42:66–71.
356. ———. 1967. Ecological aspects of the food habits of insectivorous bats. *Proc. Western Foundation Vertebrate Zool.* 1:204–63.
357. Roth, E. L. 1970. Silver-haired bat at Wichita Falls, Texas. *Southwestern Nat.* 14:449–50.
358. ———. 1972. Late Pleistocene mammals from Klein Cave, Kerr County, Texas. *Tex J. Sci.* 24:75–84.
359. Sanborn, C. C. 1932. The bats of the genus *Eumops*. *J. Mamm.* 13:347–57.
360. Scarbrough, D. L. 1989. Big free-tailed bat, *Tadarida macrotis* (Gray, 1839) from Brazos County, Texas. *Tex. J. Sci.* 41:109.
361. Schmidly, D. J. 1984. Texas mammals: diversity and geographic distribution.

Pp. 13–25. In *Protection of Texas natural diversity: an introduction for natural resource planners and managers*, ed. E. G. Carls and J. Neal. College Station: Texas Agricultural Exp. Sta., Texas A&M Univ. 60 pp.

362. ———, and F. S. Hendricks. 1984. Mammals of the San Carlos Mountains of Tamaulipas, Mexico. Pp. 15–69. In *Contributions in mammalogy in honor of Robert L. Packard*, ed. R. E. Martin and B. R. Chapman. *Spec. Publ. Mus. Texas Tech Univ.* 22:1–234.
363. ———, K. T. Wilkins, R. L. Honeycutt, and B. C. Weynand. 1977. The bats of east Texas. *Tex. J. Sci.* 28:127–43.
364. Schultz, J. G., C. D. Fisher, and S. Hightower. 1975. Recent records of the eastern big-eared bat (*Plecotus rafinesquii*) in eastern Texas. *Southwestern Nat.* 20:144–45.
365. Schwartz, A. 1955. The status of the species of the *brasiliensis* group of the genus *Tadarida. J. Mamm.* 36:106–109.
366. Selander, R. K., and J. K. Baker. 1957. The cave swallow in Texas. *Condor.* 59:345–63.
367. Semken, H. A., Jr. 1961. Fossil vertebrates from Longhorn Cavern, Burnett County, Texas. *Tex. J. Sci.* 13:290–310.
368. Shamel, H. H. 1931. Notes on the American bats of the genus *Tadarida. Proc. U.S. Natl. Mus.* 78:1–27.
369. Sherman, H. B. 1937. Breeding habits of the free-tailed bat. *J. Mamm.* 18:176–87.
370. Short, H. L. 1961a. Age at sexual maturity of Mexican free-tailed bats. *J. Mamm.* 42:533–36.
371. ———. 1961b. Growth and development of Mexican free-tailed bats. *Southwestern Nat.* 6:156–63.
372. ———, R. B. Davis, and C. F. Herreid, II. 1960. Movements of the Mexican free-tailed bat in Texas. *Southwestern Nat.* 5:208–16.
373. Shull, A. J. 1988. Endangered and threatened wildlife and plants; determination of endangered status for two long-nosed bats. *Federal Register* 53: 38456–60.
374. Slaughter, B. H., and W. L. McClure. 1965. The Sims Bayou local fauna: Pleistocene of Houston, Texas. *Tex. J. Sci.* 17:404–17.
375. Smith, A. R., and J. R. Reddell, eds. 1965. The caves of Kinney County. *Tex. Speleol. Surv.* 2:1–34.
376. Smith, D. D. 1975. Record of the red bat in Brewster County, Texas. *Tex. J. Sci.* 26:601–602.
377. Smith, J. D. 1972. Systematics of the Chiropteran family Mormoopidae. *Univ. Kansas Mus. Nat. Hist. Misc. Publ.* 56:1–132.
378. Spencer, S. G., P. C. Choucair, and B. R. Chapman. 1988. Northward expansion of the southern yellow bat, *Lasiurus ega*, in Texas. *Southwestern Nat.* 33:493.
379. Spenrath, C. A., and R. K. LaVal. 1974. An ecological study of a resident population of *Tadarida brasiliensis* in eastern Texas. *Occas. Pap. Mus. Texas Tech Univ.* 21:1–14.
380. Sprunt, A., Jr. 1950. Hawk predation at the bat caves of Texas. *Tex. J. Sci.* 2:463–70.

381. Stager, K. E. 1941. A group of bat-eating duck hawks. *Condor* 43:137–39.
382. ———. 1942. A new free-tailed bat from Texas. *Bull. Southern California Acad. Sci.* 41:49–50.
383. ———. 1948. Falcons prey on Ney Cave bats. *Nat. Speleol. Soc. Bull.* 10:49–50.
384. Stains, H. J. 1957. A new bat (genus *Leptonycteris*) from Coahuila. *Univ. Kansas Publ. Mus. Nat. Hist.* 9:353–56.
385. Stallcup, W. B. 1956. Notes on mammals of Dallas County, Texas. *Field and Laboratory.* 24:96–101.
386. Storer, T. I. 1926. Bats, bat towers and mosquitoes. *J. Mamm.* 7:85–91.
387. Straney, D. O., M. H. Smith, R. J. Baker, and I. F. Greenbaum. 1976. Biochemical variation and genic similarity of *Myotis velifer* and *Macrotus californicus. Comp. Biochem. Physiol.* 54B:243–48.
388. Strecker, J. K. 1910. Notes on the fauna of northwestern Texas. *Baylor Univ. Bull.* 18:1–31.
389. ———. 1924. The mammals of McLennan County, Texas. *Baylor Bull.* 27:1–20.
390. Struhsaker, T. T. 1961. Morphological factors regulating flight in bats. *J. Mamm.* 42:152–59.
391. Sullivan, T. D., J. E. Grimes, R. B. Eads, G. C. Menzies, and J. V. Irons. 1954. Recovery of rabies virus from colonial, insectivorous bats in Texas. *Public Health Rpts.* 69:766–68.
392. Svoboda, P. L., J. R. Choate, and R. K. Chesser. 1985. Genetic relationships among southwestern populations of the Brazilian free-tailed bat. *J. Mamm.* 66:444–50.
393. Tamsitt, J. R. 1954. The mammals of two areas in the Big Bend region of Trans-Pecos Texas. *Tex. J. Sci.* 1:33–61.
394. Taylor, W. P. 1940. Ecological classification of the mammals and birds of Walker County, Texas, and some adjoining areas. *Proc. North American Wildl. Conf.* 5:170–76.
395. Thomas, O. 1884. On the small mammals of Duval County, Texas. *Proc. Zool. Soc. London* 443–50.
396. Tibbets, T. 1956. Homing instincts of two bats, *Eptesicus fuscus* and *Tadarida mexicana* (Mammalia: Chiroptera). *Southwestern Nat.* 1:194.
397. Tinkle, D. W., and W. W. Milstead. 1960. Sex ratios and population density in hibernating *Myotis. Am. Midland Nat.* 63:327–34.
398. ———, and I. G. Patterson. 1965. A study of hibernating populations of *Myotis velifer* in northwestern Texas. *J. Mamm.* 46:612–33.
399. Tuttle, M. D., and L. R. Heaney. 1974. Maternity habits of *Myotis leibii* in South Dakota. *Bull. Southern California Acad. Sci.* 73:80–83.
400. Twente, J. W., Jr. 1954. Habitat selection of cavern-dwelling bats as illustrated by four vespertilionids. Ph.D. dissertation, Univ. of Michigan, Ann Arbor. 163 pp.
401. ———. 1955a. Some aspects of habitat selection and other behavior of cavern-dwelling bats. *Ecology* 36:706–32.
402. ———. 1955b. Aspects of a population study of cavern-dwelling bats. *J. Mamm.* 36:379–90.
403. Van Devender, T. R., G. L. Bradley, and A. H. Harris. 1987. Late Quater-

nary mammals from the Hueco Mountains, El Paso and Hudspeth counties, Texas. *Southwestern Nat.* 32:179–95.

404. van Zyll de Jong, C. G. 1979. Distribution and systematic relationships of long-eared *Myotis* in western Canada. *Canadian J. Zool.* 57:987–94.
405. ———. 1984. Taxonomic relationships of Nearctic small-footed bats of the *Myotis leibii* group (Chiroptera: vespertilionidae). *Canadian J. Zool.* 62: 2519–26.
406. Vaughan, T. A. 1959. Functional morphology of three bats: *Eumops, Myotis, Macrotus. Univ. Kansas Publ. Mus. Nat. Hist.* 12:1–153.
407. Villa-R., B. 1956. *Tadarida brasiliensis mexicana* (Sassure), el murcielago guanero, es una subspecie migratoria. *Acta Zoologica Mexicana.* 1:1–11.
408. ———, and E. L. Cockrum. 1962. Migration in the guano bat *Tadarida brasiliensis mexicana* (Sassure). *J. Mamm.* 43:43–64.
409. Walton, D. W., and J. D. Kimbrough. 1970. *Eumops perotis* from Black Gap Wildlife Refuge. *Southwestern Nat.* 15:131–43.
410. ———, and N. J. Siegel. 1966. The histology of the pararhinal glands of the pallid bat, *Antrozous pallidus. J. Mamm.* 47:357–60.
411. Warner, J. W., J. L. Patton, A. L. Gardner, and R. J. Baker. 1974. Karyotypic analysis of twenty-one species of molossid bats (Molossidae: Chiroptera). *Canadian J. Genet. Cytol.* 16:165–76.
412. Watkins, S. 1956. The old man of Frio Cave. *Tex. Game and Fish* 14:14, 26–27.
413. Werner, H. J. 1966. Observations of the facial glands of the guano bat *Tadarida brasiliensis mexicana* (Sassure). *Proc. Louisiana Acad. Sci.* 29:156–60.
414. Whitaker, J. O., Jr., and D. A. Easterla. 1975. Ectoparasites of bats from Big Bend National Park, Texas. *Southwestern Nat.* 20:241–54.
415. ———, ———, and A. Fain. 1987. First record of the mite, *Ewingana (Doreyana) doreyae* Dusbabek 1968 (Acarina, Myobiidae) from the United States with notes on streblid flies from Big Bend National Park, Texas. *Southwestern Nat.* 32:505.
416. White, P. J. 1948a. Caves of central Texas. *Nat. Speleol. Soc. Bull.* 10:46–64.
417. ———. 1948b. The Devil's sinkhole. *Nat. Speleol. Soc. Bul.* 10:3–14.
418. Wilkins, K. T., W. J. Boeer, D. S. Rogers, and W. S. Moodi. 1979. Records for eight Texas mammals. *Florida Sci.* 42:59–60.
419. Wilks, B. J., and H. E. Laughlin. 1961. Roadrunner preys on a bat. *J. Mamm.* 42:98.
420. Williams, T. C., L. C. Ireland, and J. M. Williams. 1973. High altitude flights of the free-tailed bat, *Tadarida brasiliensis*, observed with radar. *J. Mamm.* 54:807–21.
421. Wilson, D. E. 1979. Reproductive patterns. Pp. 317–78. In *Biology of bats of the New World family Phyllostomatidae*, Part III, ed. R. J. Baker, J. K. Jones, Jr., and D. C. Carter. *Spec. Publ. Mus. Texas Tech Univ.* 16:1–441.
422. ———. 1985. Status report: *Leptonycteris nivalis (Sassure), Mexican long-nosed bat.* Unpbl. report prepared for Office of Endangered Species, U.S. Fish and Wildl. Serv. 33 pp.
423. Wimsatt, W. A. 1945. Notes on breeding behavior, pregnancy, and parturi-

tion in some vespertilionid bats of the eastern United States. *J. Mamm.* 26:23–33.

424. Wiseman J. S. 1963. Predation by the Texas rat snake on the hoary bat. *J. Mamm.* 44:581.
425. ———, B. L. Davis, and J. E. Grimes. 1962. Rabies infection in the red bat, *Lasiurus borealis borealis* (Müller), in Texas. *J. Mamm.* 43:279–80.
426. Yates, T. L., W. R. Barber, and D. M. Armstrong. 1987. Survey of North American collections of Recent mammals. Supplement to *J. Mamm.* 68:1–76.
427. Young, D. B., and J. F. Scudday. 1975. An incidence of winter activity in *Myotis californicus*. *Southwestern Nat.* 19:452.
428. Zehner, W. 1985. First record of *Pipistrellus subflavus* (Chiroptera: vespertilionidae) on Padre Island, Texas. *Southwestern Nat.* 30:306.

Index

Page numbers of table and figure legends are in *italics*.

The Bats of Texas was composed into type on a Compugraphic digital phototypesetter in eleven point Goudy Old Style with two points of spacing between the lines. Goudy Old Style was also selected for display. The book was designed by Susan Pearce, typeset by Metricomp, Inc., and printed offset by Hart Graphics, Inc. The paperback books were bound by Hart Graphics, Inc. The cloth bound books were bound by Universal Bookbindery, Inc. The paper on which this book is printed carries acid-free characteristics for an effective life of at least three hundred years.

TEXAS A&M UNIVERSITY PRESS : COLLEGE STATION